William M. Gabb

Description of a collection of fossils made by Doctor Antonion Raimondi in Peru

William M. Gabb

Description of a collection of fossils made by Doctor Antonion Raimondi in Peru

ISBN/EAN: 9783742857422

Manufactured in Europe, USA, Canada, Australia, Japa

Cover: Foto ©berggeist007 / pixelio.de

Manufactured and distributed by brebook publishing software
(www.brebook.com)

William M. Gabb

Description of a collection of fossils made by Doctor Antonion Raimondi in Peru

Description of a Collection of Fossils, made by Doctor Antonio Raimondi in Peru.

By Wm. M. Gabb.

SEVERAL years ago Dr. Antonio Raimondi, of Lima, Peru, sent me a fine series of fossils collected by himself during a detailed study of that Republic, and which he assured me, in the accompanying letter, had extended over a period of eighteen years. He desired me to study and describe them; but almost uninterrupted professional engagements have never allowed me time to take them in hand. In the Proceedings of the California Academy of Natural Sciences, 1867, p. 359, I published a translation of a part of Dr. Raimondi's letter, giving a hasty sketch of the geology of the country; and in the American Journal of Conchology, vol. v, p. 25, I published what was intended as the first of a series of papers on the subject; then describing the Tertiary fossils. Since then I have never had the opportunity of resuming the work until now.

So little has been written on South American palæontology that it may be advisable to give here a list of the papers, scattered as they are in books of travel and periodicals, American and European. The following list may not be perfect, but includes all the works I have consulted, and in fact all with which I am acquainted, bearing on the question. I have appended the list of papers, together with as complete a review as possible of the fossils, as an appendix to this memoir, and trust that this work, not the most agreeable in its character, may be of use to those who follow me. In revising the species, I have compared each in a genus with all of its congeners, and have endeavored, so far as the information was available, to revise the generic references. Without specimens, this is of course always unsatisfactory, and, while I have been able to be positive in many cases, there are many others where it was impossible to be accurate. I have been fortunate enough to obtain access to the types of Dr. Isaac Lea's paper in the Transactions of the American Philosophical Society, and thereby correct several errors into which subsequent students have fallen.

PART I. TERTIARY FOSSILS.

The Tertiary fossils were described, as before said, in the American Journal of Conchology. Besides the species mentioned in that paper, there were others, too imperfect for positive determination, except the two following:—

SEMELE SOLIDA, Gray; a cast.
OSTREA IRIDESCENS, Gray; a wide-spread species.

The following forms are here figured for the first time:—
FUSUS PAYTENSIS, G., Pl. 35, fig. 1, 1a; Jour. Conch. v. 4, p. 25.
TRITONIUM PERNODOSUM, G., Pl. 35, f. 2; Jour. Conch., v. 4, p. 26.
EUSPIRA ORTONI, Gabb, Pl. 35, fig. 3.
Ampullina Ortoni, G., Jour. Conch., v. 4, p. 27.
CERITHIUM LÆVIUSCULUM, G., Pl. 35, f. 4; Jour. Conch., v. 4, p. 27.
LITTORINA LAQUEATA, G., Pl. 35, f. 5; Jour. Conch., v. 4, p. 28.
VOLUTODERMA PLICIFERA, Gabb, Pl. 35, fig. 6.
Volutilithes id., Gabb, Jour. Conch., vol. 4, p. 28.
TURRITELLA COCHLEIFORMIS, G., Pl. 35, f. 7; Jour. Conch. v. 4, p. 29.
RÆTA GIBBOSA, G., Pl. 35, fig. 8, 8a; Jour. Conch., v. 4, p. 30.
CARDIUM (LÆVICARDIUM) PERTENUE, G., Pl. 35, f. 9, 9a; Jour. Conch., v. 4, p. 30.
ARCA (SCAPHARCA) RAIMONDII, G., Pl. 35, f. 10, 10a; Jour. Conch., v. 4, p. 31.

PART II. SECONDARY FOSSILS.

In consequence of the scanty stratigraphical information accompanying the specimens, it is in some cases impossible to assign the species to their proper geological horizon with any degree of certainty. I have therefore united all of the undoubted secondary forms in a consecutive series, quoting with each species such information as I possess. Dr. Raimondi gives the altitude at which each was found, and I have in most cases added it, although not a matter of very great interest.

CEPHALOPODA.

AMMONITES, Brug.

A. ATTENUATUS, Hyatt, Pl. 36, f. 1, 1a, 1b.
Buchiceras attenuatum, Hyatt, Proc. Bost. N. H. Soc., v. 17, p. 372.

Shell flattened on the sides and back; whorls increasing somewhat rapidly in size; nodose and marked by obsolete radiating ribs. In the young shell these ribs start on the inner edge of the whorl, in contact with the preceding volution, cross the umbilical margin and develop into a rather sharp tubercle. From this tubercle they run entire, or branch into two or even three, each ending on the dorsal margin in a small tubercle. As the shell increases in size, the umbilical row of tubercles changes in character, they becoming fewer and much more prominent, and the sides of the shell more convex. Umbilicus nearly a fourth of the diameter of the shell, its margins rounding off with a regular curve into the sides between the nodes. Dorsum* flattened, rendered concave by the two dorsal rows of tubercles which are placed alternately. Aperture subovate, deeply notched by the preceding whorl. Septum comparatively simple, consisting of a dorsal and eight lateral lobes, with their corresponding saddles. The lobes are all small; the dorsal is short and broad, ending in two short branches, with all the teeth of nearly the same size, as are those on the sides of the lobe; the middle space between the branches is about one third of the depth of the lobe, and of an equal width. The entire lobe lies in the space on the dorsal surface between the tubercles. The superior lateral lobe is not more than half the length of the dorsal, very constricted above, and is expanded below into a spathalate form, bordered by six or seven teeth nearly of the same size. The dorsal saddle is divided by a deep notch into two unequal parts, the upper side indented by four teeth, the lower by but one. The first two lateral lobes are, like all the others, of the same shape as the superior lateral, but of twice the size, and end in five bidentate processes. The second and third saddles are notched by one tooth each, the fourth by two, and the others by but one each. The other lobes are of the same general pattern as already described, but diminish regularly in size and become correspondingly more simple in detail, except one on the margin of the umbilicus, which is of the same size as the third lateral.

Greatest diameter, 4.5 inches; greatest width of whorl, 1.7 in.; width of aperture, 1.5 in.; width of umbilicus, 1.0 in.

Locality. Quebrada de Huari, between Morococho and Jauja. "From a bluish calcareous sandstone extensively developed in the Cordillera of the central part of Peru. Jurassic. Height of 3300 metres above sea level." Attached to it in the same block of matrix is a fragment of *Neithea 5-costata*, and a small *Exogyra* (young of *E. polygona?*). The presence of the former shell proves it to be middle Cretaceous.

The present species is of the type of *A. Michelinianus*, d'Orbigny, of the French

---

* I use the term dorsum, in accordance with the old usage, for the peripheral portion of the shell, although recent research seems to point to this having really corresponded with the abdominal aspect of the animal. In doing so, I have no other defence to offer than that this entire paper has been written some time, and before the publication containing these new views reached me.

Gault, but has a very different septum; the tubercles on the dorsal angles are more numerous and alternate instead of being opposite, and on the sides the tubercles are fewer and more isolated. It also has fewer ribs. I sent this shell, with several other doubtful Ammonites, to Prof. Alph. Hyatt, who has named a number of species of the genus from South America. His descriptions are, however, so unsatisfactory that, as in this and the other cases where I have his species, I was unable to identify them. For this reason, therefore, I do not hesitate in redescribing the species and figuring it, so as to enable those who follow us to better recognize it. His note accompanying the specimen is as follows: "This is my *Buchiceras attenuatum* from Celedin, Peru, but the specimens described by me were smaller and younger." There are some strongly-marked points of difference between my specimen and the few characters pointed out by Prof. Hyatt, but he seems to consider them the results of difference of age.

In not following Hyatt, Waagen, and others in the generic subdivisions of the *Ammonites*, I must admit that I am not yet convinced of the utility of the proposed genera. In the present state of zoological classification it has become clear that trenchant generic, or even specific, lines cannot be drawn; in other words, with full series of specimens, genera and species resolve themselves into convenient groups, between which transitional forms occur. A genus can no longer be considered "a group of species, having a series of characters in common, and by which it can be distinguished from every other group." Rather, it is a group of species, most nearly allied to some one typical form—in short, a pure matter of convenience—and since ideas can only be expressed by words, these groups must have names which, for convenience, we call generic names. Now, in the *Ammonites* it may be very well to group the species together, and sections in the genus, whether known as "sections," "sub-genera," "groups," or what not, are undoubtedly useful; but it seems to me unnecessary to burden our already ponderous nomenclature with perhaps a hundred more names, when one will suffice. The transitions from one of the new genera to another are so gradual, so minute and perfect, that they unnecessarily increase the labor of the student, instead of simplifying it; and the prime object of classification is to obviate this. The best proof of this last assertion is that no two of the leaders of the new system agree on the course to pursue, or on the limits of the genera proposed. I might add much with reference to the genera of the *Ammonitidæ*, based on the manner of coiling. It is generally recognized that these genera are purely artificial, and that species are well known, especially in Europe, belonging at the same time to several genera. But this subject is foreign to the present paper, and I know that it is in better hands than my own. I trust that before long the studies of Prof. Hyatt on this question will be published.

A. *Sp. indet.* Pl. 36, f. 2.

A single specimen, too imperfect for detailed description. The shell is allied to *A. Renauxianus*, d'Orb., and *A. cultratus*, d'Orb. of the Neocomien. It is more strongly carinated on the dorsum than the former, and the ribs, instead of being bifurcated from a tubercle as in that species, are large and single. In this last respect, as well as in the rib being more acute, it differs from the latter.

Locality. From the "Cerro de Potosi." From the calcareous sandstone which overlies the limestone of the Cerro of San Antonio; near the silver mines; height 4200 metres.

The geological age is not stated, but the type of the species indicates either the jurassic, or a horizon very low in the Cretaceous.

A. AEGOCEROS? Phil. Pl. 36, fig. 3, 3a, 3b.
A. aegoceros, Phil., Viaje Atacama, p. 142, pl. 2, f. 2, 3.

Shell many whorled; whorls regularly rounded on the back and sides; umbilicus wide; aperture subcircular. Surface marked by numerous, small, acute ribs, which begin on the inner margin, and cross the dorsum; these ribs arch slightly forward on the middle of the side and on the dorsum. Most of them bifurcate in the middle, though an occasional one, placed at irregular distances, remains entire. In some cases, instead of branching, there is a secondary rib interpolated, its inner end not uniting with that on either side. Septum unknown.

Locality. On the label, with a corresponding number, one specimen is referred to the Hacienda of Macanga, Prov. of Pataz; but the accompanying catalogue, as well as the label on another specimen with the same number, gives "between the Pueblos of Huandoval and Corongo, Prov. of Conchucos, jurassic." The latter is more probably correct, since there are other fossils from the former locality, and their lithological character is entirely different.

Remarks. The species is nearest to *A. annulatus*, Sby. of the Lias, and *A. biplex*, Sby. of the Oxford. From the former it differs in the ribs being fewer, larger, and in being arched, instead of straight, and in more of them branching. The cross section of the whorl is also much more nearly circular. In this last respect, it also differs from the latter, as well as in having the ribs smaller and more numerous. I have referred the species to Philippi's name doubtfully, since the ribs on his figure seem to be straighter, but the figure is evidently badly drawn, and the difference is too slight to warrant a separation.

A. Hyatti, n. s., Pl. 37, fig. 1, 1a.

Shell sub-globose, whorls increasing rapidly in size, rounded; umbilicus small, umbilical face of the whorl vertical, margin rounded, sides very slightly converging, dorsum regularly rounded; surface smooth, or only marked by indistinct lines of growth, which bend forward on the inner margin of the whorl, slightly backwards on the middle, and then forward towards the dorsum. Septum unknown.

Measurements. Diameter, 4.5 in.; diameter of umbilicus, 0.7 in.; height of body whorl, 2.1 in.; width of aperture, 1.75 in.

Locality. "Near Canibamba, Prov. of Huamachuco. The rock contains coal. Height 3500 metres."

Remarks. Closely allied to *A. Sutherlandiæ*, Murch., but differs in not narrowing so rapidly towards the dorsum, and in the absence of ribs on the young shell. The surface of the only specimen is well preserved, but the thin shell will not separate from the interior to expose the septum.

A. Raimondianus, n. s., Pl. 37, f. 2, 2a.

Shell varying greatly between the young and the adult form; flattened discoidal, broad on the dorsum; sides nearly parallel; whorls increasing very gradually in size, enveloping about a third of their width; umbilicus broad, open. Young shell strongly ribbed; ribs of the same size as the interspaces, showing a tendency to be slightly tuberculated on the umbilical angle, beginning at the extreme inner edge of the whorl, in contact with the preceding volution, passing vertically out of the umbilicus, they are inclined very slightly forwards on the sides of the whorl, and, on approaching the dorsum, bend with a short curve strongly forward, producing a slightly rounded, acute angle on the median dorsal line. As the shell grows older the ribs disappear, when it acquires a diameter of about four inches. On the largest specimen, 9 inches in diameter, they are represented by faint undulations ending in an obsolete tubercle on the angle of the umbilicus. The dorsal tongue is also broadly rounded instead of being subangular.

Measurements. Diameter, 9 inches; diameter of umbilicus, 3.5 in.; width of whorl, 3.5 in.; width of aperture, 1.9 in.

Locality. "Cerro del Salto del Frayle (or friar's leap), near Chorillo, 3 leagues south of Lima. Same rock as that on the Island of San Lorenzo. But a few metres above the sea." Dr. Raimondi considers this lias.

Remarks. Of the type of *A. cymodoce*, d'Orb., from the Corralline and Kimmeridge, but differs in having smaller ribs, fewer whorls, and in the strong flexure of the ribs, and lines of growth on the dorsum.

A. CONNIFERUS, n. s., Pl. 39, fig. 1, 1a.

Shell flattened and converging on the sides, and flattened and grooved on the dorsum; whorls enveloping nearly one-half; umbilicus moderate; aperture nearly twice as broad at its widest part (across the dorsum of the enveloped whorl) as at the top. Surface marked by dichotomous ribs, which run from the umbilical margin and branch about the middle of the side. At the point of division, there are tubercles on some specimens, which are, however, absent on others. From this point they arch forward with a slight and regular curve, bearing (or not, according to variation) a slight tubercle on the dorsal margin. In one case, between each pair of bifurcated ribs, there is a supplementary rib interpolated, beginning on the line of division, and continuing to the margin, in all respects like the branching rib. From the mould in the matrix, it can be seen that the ribs were acute, and the tubercles were armed with spines. In one specimen there is no trace of ribs crossing the dorsal groove; in another it is crossed by obsolete ribs, while in a third they are as well developed here as on the sides. Septum unknown, beyond the fact that it consists of a dorsal lobe, two large ones on the sides, a large ventral, and some small details about the umbilical angle.

Measurements. Greatest diameter, 3.0 in.; width of body whorl, 1.4 in. (approximate); greatest width of aperture, 1.1 in.; width of the aperture at the dorsum, 0.6 in.

Locality. "Five leagues southeast of the village of Recuay, Dept. of Huaraz. Height 3500 metres. Jurassic."

Remarks. Three specimens imbedded in calcareous nodules, from which it is impossible to extract them, so as to show satisfactorily the umbilical region. The species is nearest allied to *A. Garantianus*, d'Orb., of the Lower Oxford, but differs in having a broad, regularly concave groove on the dorsum, and in the shape of the cross section of the whorl. In one shell the greatest width is below the middle of the aperture, above which there is a rapid convergence of the sides. There are also other minor differences in the details of ornament, especially in the presence of spines, and the sharper ribs in one shell.

A. CARBONARIUS, n. s., Pl. 38, fig. 2, 2a, 2b.

Shell large, flattened discoidal, slightly convex on the sides, converging rounded towards the dorsum, which is acute; whorls deeply enveloping; umbilicus small. There is very little difference in general appearance between the young and adult shell, the characteristic markings on a shell of 0.75 inch in diameter being proportionately stronger, but exactly like one of nearly 8 inches across. Surface marked by numerous regular ribs, slightly flattened on their upper surface, and with concave interspaces nearly as large as the ribs. These begin at the suture, inside of the umbilicus, are inclined slightly forward, flexed, and near the dorsum are bent forward with a broad gentle curve, terminating against the sharp dorsal carina. Aperture elongate ovate; septum unknown. The greatest change that occurs with age in this species is in the dorsal carina. As the shell grows older, this becomes higher and narrower until it assumes the character of a prominent plate. The ribs disappear as they reach its base. See Figure 2 b.

Measurements. From a small specimen. Greatest diameter, 2.9 in.; length of aperture, 1.4 in.; width of aperture, 0.45 in.; height of aperture above the dorsum of preceding volution, 1.0 in.

Locality. From the Liassic (?) coal mine of Pariatambo, associated with *Myophoria*, and other characteristic forms. Also from the "Cerro de la Ventanilla," at a height of 5000 metres.

Remarks. From *A. primordialis*, Schlot., of the Upper Lias, this shell differs in being less acute on the dorsum, in having a smaller umbilicus, and more flexed ribs. It is more convex than *A. complanatus*, Brug., of the same formation, and the ribs want the backward flexure of that species. In both these respects it also differs from *A. discoides*, Zict., also an Upper Lias shell. The surface markings are not unlike *A.* (?) *Peruvianus*, Von Buch, a Cretaceous fossil, probably not an Ammonite, since the whorls do not seem to envelop their predecessors. The present species is deeply enveloping. In Von Buch's figure, too, the ribs are represented as twice as wide as those of our fossil.

A. BILOBATUS, Hyatt, Pl. 38, fig. 3, 3a, 3b.
Buchiceras bilobatum, Hyatt, Proc. Bost. N. H. Soc. vol. 17, p. 370.

Shell broad, robust, whorls increasing rapidly in size, and deeply enveloping; umbilicus deep, less than a third of the diameter of the shell; sides flattened and slightly converging; dorsum broad, flattened, and bearing a broad, but not high, median ridge; aperture nearly square with the corners rounded and the dorsal face a trifle the narrowest. Surface marked by a few very large ribs beginning just inside of the umbilical angle, slightly tuberculated on the angle and running to the dorsal margin, where they carry another small angular tubercle. An occasional supplementary rib is interpolated on the dorsal half of the side. Septum consisting of a dorsal lobe, two laterals on the side of the whorl, and two on the umbilical face. Dorsal lobe broad and short, and lying between the two rows of tubercles; at the corners it ends in two short slender branches, simple on their inner face and tridentate outside; between them is a broad emargination, its base nearly straight, and bearing six small teeth; above the branches are two large teeth on each side. Dorsal saddle broad, almost without indentations, except a tongue which divides it into two unequal parts, the larger being on the dorsal side, the smaller is unsymmetrical, sloping down obliquely to the superior lateral lobe, which is broad, short, and unsymmetrical. At its termination it is divided into two branches, of which the upper is narrow and obscurely tridentate; the lower is divided into one simple and one bifurcate prong, all of the same length; above the terminal branches the lobe bears a large tooth on the upper side, and a small one on the lower. The lateral saddle is a repetition, on a smaller scale, of the preceding. The inferior lateral lobe is not more than half the size of the superior, and shows only the rudiments of the same details, in the shape of three minute teeth on the end. The inferior lateral saddle, bending round the umbilical margin, is straight above, divided like the others into two unequal parts, of which the largest is on the upper side, but, instead of the tongue, the division is produced by two little teeth; the upper part is further subdivided by another tooth. The first lobe inside of the umbilicus is a mere bidentate tongue, with a little tooth on the upper side; the second lobe is still more rudimentary, and is partly hidden by the suture; the included saddle is irregularly divided into three parts by two teeth.

Measurements. Greatest diameter, 3.0 in.; diameter of the umbilicus, 0.8 in.; depth of umbilicus from the inner angle of the mouth, 0.8 in.; width of body whorl, 1.4 in.; greatest width of mouth, 1.5 in.; width of mouth at the dorsal side, 1.1 in.; depth of emargination of the ventral side of the mouth, 0.3 in.

Locality. From the "Quebrada de Colpamayo, in the immediate neighborhood of Chota, Dept. of Cajamarca. Height 2000 metres. Cretaceous?"

Remarks. From the type of the shell, I agree with Dr. Raimondi in believing this to be a Cretaceous species. Its septum is remarkably simple, but, at the same time, its simplicity has nothing that would suggest or approach to *Ceratites*. It is rather that usually observed in a young *Ammonite* that has not yet developed its details, although my specimen, as will be observed above, is 3 inches in diameter. Prof. Hyatt's note, on returning me the specimen, is: "This is undoubtedly my *Buchiceras bilobatum* from Cachiyacu, but is an older specimen than that described by me."

A. OLLONENSIS, n. s., Pl. 38, fig. 4, 4a.

Shell discoidal, compressed, umbilicus small, shallow; sides converging with a gentle curve; dorsum narrow, slightly concave. Sides slightly undulated by very broad rudimentary ribs; a row of small tubercles on the umbilical margin, and numerous small compressed tubercles bordering the dorsal margin, which are thereby rendered subacute and undulated. Aperture long and narrow, widest in the middle. Septum composed of numerous small lobes and saddles, apparently about six lobes on the sides. The dorsal (or as it might perhaps more properly be called, the abdominal lobe) is small, and lies entirely on the dorsal surface; the superior lateral lobe is much smaller than the second, from which the others diminish regularly in size. The saddles are all comparatively simple except the first, which is unequally divided.

Measurements. Greatest diameter (apparently), about 3.5 in.; greatest width of body whorl, about 1.75 in.; greatest width of aperture, 1.0 in.

Locality. "Immediate vicinity of Ollon, Prov. of Cajatambo; height 3000 metres. This rock appears to be Cretaceous." (R.)

Remarks. Described from two fragments, one consisting of the body chamber only, showing the broken remains of the first septum, a trace of the umbilical margin, but a very good surface, although somewhat broken about the lower part of the mouth. The species may be compared with *A. placenta*, DeKay, of the Cretaceous of the United States, from the flat-backed variety of which it differs in the dorsum being much wider, and bordered by tubercles instead of a smooth keel. The ventral row of tubercles, of which traces remain in our specimen, is also wanting in that species. From *A. Pedernalis*, Von Buch, which is sometimes undulated on the sides, it also differs in all the above characters, and in the ventral half of the sides not sloping down so gently to the umbilical margin. Prof. Hyatt returned me the specimen figured, with the note following: "This I should consider a variety of *Buchiceras attenuatum*, something between the form described by me and that belonging to you *(see figure of A. attenuatus,* W. M. G.), see sutures. It may be different, however, since I think I have seen Texas specimens like it, and smooth throughout, A. H." I hardly feel warranted, however, in accepting my friend's idea in this case, and venture to name the form. My reasons for separating it are, from the (imperfect, it is true) view of the septum, obtainable on the broken face, there seem to be fewer lobes than in Hyatt's species. This specimen is nearly as large as the other, and it has developed in a markedly different manner; the entire cross section of the whorl differs; there is no trace of tubercles on the sides of the volutions, and the little tubercles on the umbilical margin in this, are unrepresented in that.

A. MACROCEPHALUS, Schlot., Pl. 39, fig. 1, 1a.

Shell globose, sides and dorsum rounded; sides converging very slightly; umbilicus minute; whorls very deeply enveloped, the last almost entirely hiding the preceding; surface ornamented by small regular ribs, which arch forwards a little on the middle of the sides, and then cross the dorsum continuously. These are rarely branched, but one or even two supplementary ribs are interpolated on the dorsal half of the shell.

Measurements. Greatest diameter, 2.05 in.; greatest width of body whorl, 1.1 in.; width of mouth, 1.3 in.; height of aperture from dorsum of included whorl, 0.65 in.

Locality. "Province of Tarapaca, in the south of Peru." Jurassic. This is evidently a stray specimen, since there is no exact locality given, nor the altitude, nor yet even a guess at the geological age. This shell, also reported from Chili and Bolivia, is of especial interest, since it assists us to fix the horizon of what seems to be an important formation in the Andes.

A. ACUTISSIMUS, n. s., Pl. 36, fig. 4, 4a.

Shell discoidal, sides compressed, dorsum expanded and acute; whorls slightly convex on the sides, and marked by two rows of large rounded nodes, one near the umbilical margin, the other towards the dorsum. These nodes are placed so that the outer and inner row alternate, and they are connected with each other by broad, faint undulations. Above the outer row of nodes the sides are concave. Septum unknown.

Measurements. Width of body whorl, 1.1 in.; width of aperture, 0.5 in.

Locality. "Ridge of the three crosses, between Huallanca and Aguamiro; Prov. of Huamalies. Height more than 4000 metres. From a pulverulent carbonate of lime that appears to be Cretaceous."

Remarks. This description is from a single small fragment, so imperfect that I should not have felt warranted in describing it, were it not of so marked a form as to run little risk of being mistaken. Its knife-like dorsum and the broad nodes on the sides are characters which will separate it at a glance from any species known to me. Hyatt says of it: "If this is from the Cretaceous, it may be closely allied to my *Buchiceras serratum*, but it is not probably the same."

A. VENTANILLENSIS, n. s., Pl. 39, fig. 2, 2a-d.

Shell flattened discoidal; back broad, whorls increasing very gradually in size, but slightly enveloping; aperture subquadrate; umbilicus broad. Very young shell smooth, without ribs, sides rounded, back with a small median keel, and each whorl enveloping about half of its predecessor. At a little over an inch in diameter, the shell acquires large rounded plain ribs, the tubercles beginning to appear on the fifth or sixth rib. In the adult individuals, with the volutions nearly two inches wide, the surface characters are entirely different. The umbilical margin is rounded, and gives rise to numerous, closely placed, rounded ribs, starting with a marked inclination backwards; these afterwards curve forwards on the lower half of the whorl, from the middle continue transverse to the dorsal angle, and then incline strongly forward, forming an acute angle between the two sides against a large median dorsal keel. At the dorsal angle each rib is developed into a large flattened tubercle placed at an angle of about 45° to the corresponding one on the opposite side. Below this large tubercle, on each rib, is a smaller one, placed on the middle of the upper half of the side of the whorl. Septum unknown.

Measurements. Height of aperture, 1.5 in.; width of aperture, 1.1 in.

Localities. Liassic; one adult fragment from the "limestone of the Cerro del Ventanillo, between Pachachaca and Jauja. Height of 5000 metres." Another, and a very young shell, from the "coal mine of Pariatambo, 5 leagues from Morococho. Height 4000 metres." Dr. Raimondi considers both these localities Jurassic, and remarks of the former that the formation is very extensive. Still another specimen, a little over an inch in diameter, is from the "neighborhood of Fingo, Prov. of Huari, Dept. of Huaraz, from a schist containing coal."

Remarks. Our species is very closely allied to the Liassic form, *A. spinatus,* Brug., and may eventually prove to be identical, since they only differ in details of ornament, both going through the same series of changes from youth to the adult stage. In the specimens before me, however, the characters on which I have depended for a separation are constant. There is a broad concave space on each side of the keel, between it and the tubercles, the space between the outer sides of which is almost as great as the greatest width of the aperture. In *A. spinatus* these tubercles are placed entirely on top, and the second row, which in our shell is well down on the sides, forms the outer margin of the dorsum in that. Further, in our shell the ribs on the ventral half of the whorl are markedly flexuous, while in *A. spinatus* they are straight.

A. GIBBONIANUS, Lea, Tr. Amer. Philos. Soc., 2 S., vol. 7, p. 254, pl. 8, f. 3.

A single fragment, not more perfect than Mr. Lea's specimen, occurs in the collection from "between Huandoval and Corongo." It shows no additional characters, being a piece of one whorl, about as long as broad, and having but four ribs and part of another. It is more convex in its cross section than the form figured by Marcou. Mr. Lea's original, now in the museum of the Academy of Natural Sciences, although weathered on one side, shows that it had a rounded dorsum. The shell from Texas referred to this species by Marcou (Geol. of N. A., p. 35, pl. 2, f. 2), has an acutely carinated dorsum, and the ribs are acute, while the South American has rounded ribs. These ribs in *Gibbonianus* run all of the way to, and apparently cross the dorsum. In the Texan they stop short. The cross sections of the whorl differ markedly, though no specimen, I have seen, of *Gibbonianus* shows an entire section. Enough, however, is known to show that the umbilical margin is deep, and nearly at right angles to the side of the whorl.

A. ANDII, n. s., Pl. 39, fig. 3, 3a, 3b.

Shell discoidal, very convex in the middle and compressed on the dorsum; whorls entirely enveloping; umbilicus minute, deep; aperture broadly cordate, deeply emarginate on the ventral side. Surface marked by strong lines of growth, slightly sinuous on the sides and arching strongly forwards close to the dorsum. Besides these, on the dorsal half of the side are obsolete ribs, and near the margin a row of barely perceptible tubercles. Septum consisting of a dorsal and three lateral lobes. The dorsal lobe is large, broader above than below, and ends in a slender branch on each side, bearing one digitation on its inner side and two on the outer; above this, at the base of the lobe, is a long, narrow tridigitate branch. The adjoining saddle is very oblique, and is unequally divided by a large tongue. The superior lateral lobe is small and narrow, ending in an oblique branch bearing three teeth on its lower side and one longer process on the upper; above it, on the body of the lobe, are two processes on the upper side, and one more complex on the lower. The next saddle is oblique, higher on the lower side and indented by three teeth of unequal size. The inferior lateral lobe is small, and is divided into three bidentate branches of nearly equal size. The next saddle, occupying the curved surface adjoining the umbilicus, is nearly straight, divided in the middle by a small tongue, the upper half again subdivided by a smaller tongue. Up to this point the septa continue entirely separate, but the next lobe, lying on the umbilical face, is very broad, divides into three broad branches, some of the points of which are obliterated by abutting against the last saddle of the preceding septum.

Measurements. Greatest diameter, 2.5 inches; greatest width of aperture, 2.0 in.; greatest width of body whorl, 1.6 in.; height of aperture from the dorsum of the preceding whorl, 1.0 in.

Locality. "Province of Pataz; Dept. of Libertad; height, 3000 metres."

Remarks. In his notes, Dr. Raimondi considers the formation Cretaceous, but we have in this single fossil a sufficient proof of its being Jurassic. The specimens are in a beautiful state of preservation, and come from a light-colored, crystalline limestone showing every detail in perfection. The species belongs to the group of which *A. cordatus*, Sowerby, is the type. It resembles that shell in external character to some extent; so much so that I should have probably hesitated in separating it, had I not possessed the details of the septum. In this, it differs alike from all the known species of the group. The inferior lateral lobe is entirely unlike that of *A. cordatus*, as is all the remainder of the septum to the suture; and the resemblances in the dorsal and superior lateral lobes are very remote. From the North American *A. cordiformis*, Meek and Hayden, the differences of the septum are still greater; besides which, the only known specimen of that species is strongly costate on the surface, and much more compressed, although these external characters might be only individual in character. Prof. Hyatt, to whom I sent the specimen figured, returned it to me labelled *cordatus*, but I suspect he did not compare the septa of the two forms.

GASTEROPODA.

PERISSOLAX, Gabb.

P. TROCHIOIDES, n. s., Pl. 39, fig. 4.

Shell trochiform, spire low; body whorl convexly flattened above and sloping; below, it slopes rapidly inwards; aperture broad, irregularly rounded subquadrate; canal produced.

Locality. Hacienda of Macanga; Prov. of Pataz. Cretaceous.

Remarks. I have ventured to describe this species from a couple of casts, retaining all the parts except the long slender canal, characteristic of the genus. They retain no part of the shell, but show traces on the broadest part of the whorls of three or four revolving ribs. This species is allied to *P. longirostris*, d'Orb., sp. (*Pyrula*), Amer. Merid., p. 119, Pl. 12, fig. 13; Voy. Astrolabe and Zélée, Pl. 4, f. 30; but in that species, as well as in *P. Hombroniana*, the body whorl is broader and rounder, while in ours the greatest width is near the upper angle.

Indet. Several other casts from the same locality occur in the collection. They are from a buff-colored limestone, but none show specific characters; and, in most, even the means of determining the genus is lost.

GYRODES, Con.

G. CONTRACTA, n. s., Pl. 39, fig. 5, 5a.

Shell small, subglobose, oblique; spire low, whorls five, flattened and faintly channelled above; suture distinct and bordered by a slight thickening of the succeeding whorl; body whorl most convex in the middle, contracted in advance; umbilicus open, but unusually narrow; aperture nearly semicircular, with the inner margin vertical, and subangulated above by the flattened top of the whorl; lips simple. Surface marked by very oblique lines of growth, which are most pronounced on the upper surfaces, near and adjoining the suture. Towards the base there are a few very faint rudiments of revolving lines.

Figure. Somewhat enlarged.

Locality. Liassic coal mine of Pariatambo; and from the elevated table-land two leagues from Cajamarca.

Remarks. This shell possesses remarkable interest from its being the first species of its group of Naticas, found outside of the Upper Cretaceous. Its evidence, however, of a Cretaceous age of the deposit is counterbalanced by the presence of *Myophoria* and by the close relationship of the associated *Ammonites* to Liassic types. While possessing every character essential to the genus, of which I am acquainted with almost every described species, it differs from its congeners in having the characteristic umbilicus, but unusually narrow; an amount of difference that is not surprising when we consider the difference in geological age.

G. LIRATA, n. s., Pl. 39, fig. 6, 6a.

Shell subglobose, large; spire elevated; whorls broadly convex, but very slightly oblique, somewhat flattened on the top and regularly rounded on the sides; umbilicus patulous; aperture not very oblique. Surface marked by about five slightly raised revolving ridges crossed by well-marked lines of growth.

Figures. Natural size.

Locality. "Near Ollon, Prov. of Cajatamba. Height more than 3000 metres." Cretaceous.

Remarks. One of the largest species of the genus yet discovered. It is at once distinguished by a character, rare in the family—surface sculpture. One or two species of *Gyrodes* are crenate on the upper or lower margin of the whorls, but this has distinct revolving ridges placed at about equal distances over the whole surface. The only specimen is a cast retaining enough of the shell to show all of the specific characters except the details of surface adjoining the suture. The cast shows that the upper part of the whorl was more or less flattened, but does not show if the margin of the flattened space was acute, rounded, or crenate. The internal mould shows no trace of the surface ridges.

PRISCONATICA. N. Gen.

Among the earlier forms of Naticas (those of the Secondary rocks) is a large group that has been described, almost without exception, under the vague generic title of *Natica*. They are characterized by being almost always of large size, including several of the largest known species of the family; by their thin shells, generally elevated spire, increasing rapidly in their axial length, rather than obliquely; with very small, or entirely obsolete, umbilicus, and in having the columellar lip always thinly encrusted. In many respects they approach *Amauropsis*, but they are more naticoid in style, rarely so elongate, and never have the spiral sculpture, often found in that genus. *Ampullina*, Faujas, 1803, of which the type is *A. fluctuata*, is a heavy shell, with a peculiar incrustation of the inner lip, the callus blending gradually into the outer surface. *Ampullina*, Lam., 1813, was proposed for the thin shells, with more or less of a carina bordering the umbilical region, and which he had previously included in the fresh-water genus *Ampullaria*. For this genus the name of *Euspira*, Agas., 1837, must be used, the type being *Ampullaria sigaretina*, Min. Conch., Pl. 283, figs. 1–3. *Globularia*, Swainson, 1840, is also a synonym of this genus. Stoliczka, in Pal. Indica, incorrectly places some species of our new genus in *Ampullina*. *Natica Pedernalis*, Roemer, Kreidebildungen von Texas, may be taken as our type. A better figure than Roemer's will be found in the Palæontology of California, vol. ii. Pl. 35, f. 3.

P. OVOIDEA, n. s., Pl. 39, fig. 7.

Shell large, ovate, thin; spire moderately elevated, whorls about five, sloping above, not very convex; suture small; body whorl broadly convex, somewhat oblique; mouth sub-ovate, oblique, acute behind, broadly rounded in advance; inner lip encrusted by a thin layer. Surface marked only by lines of growth.

Measurements. Length, 3.3 in.; greatest width of body whorl, 2.4 in.; length of aperture, 2.5 in. The above measurements, except the width, are only approximate, since the extreme apex and the end of the aperture are imperfect.

Locality. Neighborhood of Ollon; from a dark bluish limestone, different from the matrix of the other Ollon specimens. Cretaceous (?).

Remarks. This species is nearest to *P. Pedernalis* in form, but is more slender and less oblique; the spire is lower than *P. prægrandis (Natica id.*, Roem.), also from the Cretaceous of Texas, and probably Northern Mexico. It is of the type "*Natica*" *Elea*, d'Orb., of the French Jura, at least so far as can be ascertained from the back view given by d'Orbigny, but its spire is much lower. There is no described species in South America approaching it in size, except *P. prælonga* (*Natica*, Seym.), from Brazil, Columbia, and France, and this has a spire as long as the body whorl.

P. INCONSPICUA, n. s., Pl. 39, fig. 8.

Shell small, elongate ovate; spire very high; whorls five and a half; rounded; upper surface rounded, sub-truncate; body whorl broadly rounded in the middle, sloping inwards below; aperture sub-circular; umbilicus absent, or very small; inner lip very thinly encrusted. Surface smooth.

Figure. Magnified to twice natural length.
Locality. Broken from the same block with the preceding.

P. AMPLA, n. s., Pl. 40, fig. 1.

Shell large, unusually short and broad, spire elevated; whorls about six (?), flattened on top, and convex in the middle; suture channelled; aperture proportionately small, oblique, equally broad above and below.

Figure. Natural size.

Locality. Between the River Chonta and the village of Baños; Prov. of Huamalies; apparently Cretaceous.

Remarks. There is but a single cast of this species in the collection, showing nothing of the surface; but the outline of the shell is so strongly characteristic, that I have not hesitated to name it. Its high spire and broad whorls are sufficient to separate it from every species I have ever encountered. The details of surface about the umbilical region are destroyed, but the cast shows enough to demonstrate that the shell was nearly imperforate, and that the inner lip was but thinly encrusted.

This is apparently the shell figured in Wilkes' Expedition Report, Pl. 15, fig. 3a, b, under the name of "*Turbo sp.*?"

TURRITELLA.

T. RAIMONDII, n. s., Pl. 40, fig. 2.

Shell long, slender, many whorled; whorls flattened on the sides, and marked by four beaded ribs, the upper one of which forms the upper margin of the whorl; between these are concave interspaces, marked by fine elevated lines; the lower angle of the body whorl is marked by a plain, sharp, linear rib, which, in the preceding volutions, forms the sutural margin. On the under surface of the last whorl there is a second plain rib of equal size, parallel with the first, and three or four smaller, between which are others still finer. Under surface of body whorl slightly concave. Outer lip broadly and sub-angularly emarginate in the middle, and produced below.

Measurements. Diameter, 0.2 in. Total length probably about 1.5 to 2.0 inch.

Locality. From the Liassic (?) coal mine of Pariatambo.

Remarks. Associated with the fragments from which this species is described, is another single piece of three volutions, which differs from the typical form in the same manner as occurs in other beaded *Turritellas*, markedly so in the case of *T. seriatim-granulata*, Roem., of the Texan Cretaceous. The entire surface is covered with coarsely beaded ribs, six in number, so closely placed as to leave no concave interspaces, and consequently no room for the finer linear markings described above.

T. PERUANA, n. s., Pl. 40, fig. 3.

Shell moderate in size, long, slender, many whorled; whorls flattened on the sides, or very slightly concave, sloping outwards below; top flattened, narrow and sloping; surface marked by three large revolving ribs, one on the upper angle, one in the middle, and the last near the base; basal margin angular; base slightly convex; outer lip deeply and roundly emarginate above, produced towards the base.

Measurements. Length of three whorls, 0.75 in.; width of body whorl, 0.25 in.

Locality. From the hacienda of Macanga, Prov. of Pataz. Cretaceous. Associated with the *Perissolax trochoides* and *Niethea quinquecostata*.

TYLOSTOMA, Sharpe, 1849.

Varigera, d'Orb., 1850.*

T. MUTABILIS, Gabb, Pl. 40, fig. 4, 4a.

T. mutabilis, Gabb, Palæontology of California, p. 261, Pl. 35, f. 6.

Shell subglobose, spire high, whorls six or seven; body whorl most convex above the middle; varices oblique, large; aperture elongate-ovate. Surface unknown.

From near the ruins of Cuelapo, twelve leagues from Chachapoyas and "hacienda of Uchupata, Province of Huari, height of 2000 metres." Cretaceous.

Remarks. Described from casts retaining almost none of the shell. Found associated with *Echinus Bolivari*, d'Orb. The specimens are all more or less distorted, but I cannot find, in the absence of surface characters, any on which to separate this from the shell described by me under the above name from Mexico.

CINULIA, Gray.

C. ANTIQUA, n. s., Pl. 40, fig. 5, 5a.

Shell ovate, convex; spire high; whorls broadly and regularly convex, outline of one side of the body whorl nearly a perfect segment of a circle; suture impressed; surface marked by about twenty regular equal rounded ribs with smaller interspaces. Aperture elongate, narrowed behind; outer lip very much thickened by a strong marginal rim; details of inner lip unknown.

Measurements. Length, 0.43 in.; greatest width, 0.25; length of aperture, 0.32 in.

* Although d'Orbigny claims the date of 1847 for this genus it cannot be allowed, since it was only indicated by name without a recognizable description in the Prodrome, and his first full description was published in 1850; while Sharpe had already characterized the genus a year previously.

Locality. Cerro del Ventanillo, between Pachachaca and Jauja, associated with *Ammonites Ventanillensis* and *Petropoma Peruanus*, both of which occur at the Liassic (?) coal mine of Pariatamba.

ACTÆONELLA, d'Orb.

A. oviformis, n. s., Pl. 40, fig. 6.

Shell thick, ovate, a little narrower below than above; spire slightly elevated; body whorl broadly and regularly convex, narrowing in advance; aperture narrow above, below broader; inner lip encrusted and with three large folds in advance, the middle of which is narrowest and highest. Surface smooth; marked only by lines of growth.

Locality. From a gray crystalline limestone in the neighborhood of Ollon. Cretaceous.

Remarks. The species is described from a somewhat imperfect specimen, the fractures of the matrix having crossed the shell also, and broken away parts. The extreme apex is hidden and most of the aperture wanting, though the cross section of the cavity gives its shape. It is nearest in form to *A. gigantea*, d'Orb., but it has a higher spire and is more convex on the sides and not so acute anteriorly.

PETROPOMA, N. Gen.

Shell trochoid; spire more or less elevated; umbilicus small or imperforate; inner lip encrusted on the body whorl and slightly thickened on the umbilical margin, not toothed; aperture subcircular. Operculum multispiral, subcircular, thick, slightly conical externally and showing the volutions; internally each volution is expanded on its inner margin so as to cover all the surface except a little central pit.

The general character of the shell is such that it could, except for the operculum, have been referred to the genus *Gibbula*, perhaps more properly coming into the sub-genus *Forskälia* of H. and A. Adams, in consequence of the flattened sides to the whorls in the species before us. But like most, if not all the other described Trochoids, that genus is characterized by a horny operculum. In ours it is fully as massive as the shell, and its obliquely truncated margins show that it probably did not possess even a corneous expansion. I have not hesitated in associating this operculum with the shell, since both are equally abundant in the rock in which they are found, and there is no other shell to which the operculum can be referred. There is no doubt but that many of the species of fossils arbitrarily referred to *Trochus* and *Turbo* will have to be separated as soon as sufficient details of their character shall have been obtained. Notable examples of this may be seen in such monographs as d'Orbigny's Pal. Française. The operculum figured in that work (Terr. Cret., vol ii. Pl. 186 *bis*, figs. 15–17) evidently belongs to this genus. The author declares himself unable to associate it with its proper shell, and says of the two operculi figured that they are "au moins un confirmation de leur bon classement dans le genre *Turbo*. L'un d'eux est très-remarquable par ses tours très-rapprochés, comme chez les *Trochus* proprement dits."

In virtue of the general form and the multispiral operculum, I believe this genus to belong near to *Gibbula*, despite the fact of the calcareous nature of the operculum. All of the genera in the *Turbinidæ* approaching it in form have that member markedly paucispiral, and constructed in an entirely different manner.

P. PERUANUS, n. s., Pl. 40, fig. 8.

Shell small, robust, spire moderately elevated, sides regularly sloping; whorls six, flattened on their upper sides; body whorl flattened and sloping above, rounded and bi- or tri-carinate on the margin, and slightly convex below. Surface marked by about six heavily beaded ribs above, the uppermost of which makes the sutural margin; the two on the sides are the largest, and when a third exists it is slightly smaller and placed below the other two; under surface covered with six or seven similar ribs, the last forming a crenulated margin to the umbilicus, which is minute and apparently imperforate. Aperture sloping subquadrate; outer lip simple; inner lip bordered by a flattened pillar lip, the anterior part slightly overlapping the umbilicus.

Locality. From the coal mine of Pariatambo and from the Cerro del Ventanillo. Liassic.

HELCION, Montf.

H. CARBONARIUS, n. s., Pl. 40, fig. 7, 7a.

Shell small, nearly circular, thin; apex subcentral, about a third as high as the diameter of the shell; surface polished and marked by prominent lines of growth.

Locality. With the preceding.

LAMELLIBRANCHIATA.

PANOPÆA, Menard.

P. UNDULATA, n. s., Pl. 40, fig. 9.

Shell sub-elliptic, moderately convex; base nearly straight in the middle; beak prominent; anterior end unknown; posterior gaping, nearly semicircular and with the upper and middle margins reflexed. Surface covered with regular, rounded concentric ribs and subacute interspaces; the ribs larger on the middle and smaller towards the base; posterior to the beaks are traces of very faint radiating lines, more closely placed than the concentric ribs and barely perceptible except in the interspaces.

Figure. Natural size.

Locality. From the Pueblo of Pion, in the department of the Amazons. Two isolated specimens without associates, considered Cretaceous by Dr. Raimondi.

Remarks. This shell can be distinguished by the rounded gaping posterior end and gently dilated margins, as well as by the marked concentric ornamentation. It is from a hard white limestone unlike any other rock among the upwards of 400 specimens in the collection.

CORBULA, Brug.

C., *sp. indet.*

Numerous casts of a small equivalve shell, doubtless of this genus, occur among the Cretaceous fossils of the neighborhood of Ollon. Beyond the fact that they are nearly equilateral and equivalve (a slightly greater convexity existing in the right valve and the characteristic biangularity posteriorly), they show no characters. I therefore do not feel warranted in naming the species, since no good diagnostic characters can be given from the material.

C. NUCULOIDES, n. s., Pl. 40, fig. 10.

Shell small, nearly equivalve; beaks very slightly in advance of the middle; anterior end elongately rounded, base regularly curved; posterior cardinal margin concave, the surface flattened; posterior end produced, narrow and rounded. Surface marked by fine lines of growth.

Locality. From the coal mine of Pariatamba. Lias.

Remarks. From the other two species, with which it is associated, this shell can be distinguished by its form, resembling a *Nuculana*, and by its surface, nearly devoid of the heavy lines so characteristic of its genus.

C. PERUANA, n. s., Pl. 40, fig. 11.

Shell small, very inequilateral; beaks placed in advance of the middle; anterior end sloping convexly above and most prominent near the base; base broadly and slightly convex; margin abruptly bent down; posterior side excavated behind the beaks, a concave area running to the posterior end and bounded by a sharp angle; posterior end narrow and produced. Surface marked by small concentric ribs.

Locality. With the preceding.

Remarks. A very characteristic *Corbula*, recognizable from even a fragment both by its outline and markings. It differs from the preceding by its more cuneiform shape and its ribs. From the following it can be distinguished by the fine ribs, smaller size, and less truncated anterior end. Associated with this is a single specimen, somewhat mutilated, so that the shape is not all retained. It has smaller ribs, and shows faint traces of radiating lines. I believe it to be different, but have not material for describing it. Both are left valves.

C. RAIMONDII, n. s., Pl. 40, fig. 13.

Shell small, robust, cuneiform; beak of right valve large, prominent, and involute; anterior end rounded, truncate, sloping nearly straight and obliquely to the base, where it is narrowly rounded. Base broadly convex; posterior side concave above, the area joining the surface by a regular curve, and not bounded by a ridge as in the preceding species; posterior end narrow and produced. Surface marked by large, rounded concentric ribs.

Locality. With the two preceding species.

Remarks. Easily distinguished by its large, coarse ribs, its prominent beak, the nearly straight sloping anterior end, and by the absence of a ridge running from the beaks to the upper part of the posterior end. The most marked species of the three.

PHOLADOMYA, Sby.

P. AUSTRALIS, n. s., Pl. 40, fig. 14.

Shell small, oblique, very inequilateral, convex; beaks placed close to the anterior end, large, incurved; anterior end narrowly rounded; posterior broad, regularly rounded, and slightly oblique. Surface smooth in advance; seven radiating ribs run from the beaks to the base, their ends occupying almost the entire basal margin; posterior to these the surface is covered with small rounded, concentric ribs, faint traces of which can be seen between the radiations; posterior to the beaks and adjoining the cardinal margin, is a small, narrow, cordate area, slightly concave and bounded by a slight angle.

Figure. Natural size.

Locality. Hacienda of Macanga. Cretaceous.

Remarks. Two suites of specimens, one from near Cajamarca, the other from the province of Huari, have the same size and general form, but show no surface markings.

P. RAIMONDII, n. s., Pl. 40, fig. 15.

Shell very inequilateral; beaks placed about a third of the length from the anterior end, which is broadly and regularly rounded; posterior end narrower than the anterior, and produced; base most prominent directly under the beaks. Surface covered by radiating ribs, which become faint at each end. A slight narrow depression runs parallel with the posterior cardinal margin, bordered by a rounded angle.

Figure. Natural size.

Locality. Between Combayoa and Polloc; Dept. of Cajamarca. From a limestone which, Dr. Raimondi says, "appears to be Jurassic." Two other specimens, referred doubtfully to the same species, are from the hacienda of Macanga, which locality is almost certainly Cretaceous.

P. ZIETENII, Agas., Etud. sur les Myes, p. 54, Pl. 3, f. 13-15.
P. *fidicula*, Zieten (not Sby.).
P. *Zieteni*, Bayle & Coquand, Mem. G. Soc. Fr., 2 s., v. 4, p. 26, Pl. 7, f. 8.

Our shell is the same as that figured by the last authors, though I think it is doubtful if this is the same as the European species; it differs in the posterior end and is much more strongly ribbed than Agassiz's figure. It is from the Jurassic(?) rocks in the neighborhood of Ollon.

P., *sp. indet.*

From the Liassic coal mines of Pariatambo there is a single small *Pholodomya*, too much distorted for a very satisfactory determination, and entirely unfitted for description if new, since the outline is destroyed. In style it resembles *P. echinata*, Agas., but has fewer radiating ribs.

HOMOMYA, Agas.

H. INCURVA, n. s., Pl. 40, fig. 13, 13a.

Shell convex, inequilateral; beaks prominent, placed about a third of the length from the anterior end, and strongly incurved; anterior end most prominent just above the middle, above which point to the umbones it is slightly concave, and below which it unites by a broad curve with the broadly and regularly rounded bases; posterior cardinal margin sinuous; posterior end obliquely rounded, subtruncate above, narrowly rounded at its most prominent part in the middle. Surface marked by numerous small rounded concentric ribs.

Figures. Natural size.

Locality. From the neighborhood of Ollon. Cretaceous or Jurassic?

Remarks. Easily distinguished by its short, compact form and its strongly incurved beak. The concentric lines vary in strength in different specimens; in some they are a tenth of an inch across, while in others they diminish to mere lines of growth. I cannot think this indicates two species, because, as is usual in such cases, the undulations are strong on the beaks, and the fine lined forms show this character on the middle or towards the base. Still, nearly all the specimens being more or less imperfect, I am not certain that there is not also a difference in outline.

II. PONDEROSA, n. s., Pl. 41, fig. 1.

Shell large; beaks prominent, placed about a third of the length from the anterior end; strongly incurved; anterior end sloping nearly straight to the middle; base broadly and pretty regularly rounded, not prominent; posterior end most prominent near the base. Surface (of cast) marked by only very faint concentric lines, and with a broad faint depression towards the base a little posterior to the middle.

Length, 4.5 inches; width, from beak to base, 3.75 in.; width of two valves, 2.5 inches.

Locality. From the elevated table land two leagues from Cajamarca. Considered Cretaceous by Dr. Raimondi.

TELLINA, Linn.

?T. PERUANA, n. s., Pl. 40, fig. 16.

Shell lenticular, broad, sides very flattened; beaks minute, placed about a third of the length from the anterior end; base and anterior end regularly rounded, posterior end somewhat produced, sloping downwards from the beaks and rounded sub-biangular. Surface marked by regular fine concentric striæ.

Figure. Natural size.

Locality. From the coal mine of Pariatambo.

Remarks. Not a rare species; represented by a number of specimens. I have referred it doubtfully to the above genus, because I have been unable to expose the hinge, or even the pallial impression.

CARDIUM, Linn.
Subgen. PROTOCARDIA.

C. (P.) APPRESSUM, n. s., Pl. 40, fig. 17.

Shell thin, not very convex, nearly equilateral, beaks central, anterior end most prominent in the middle, posterior rounded, obliquely truncated; base broadly and evenly rounded; surface marked by small, regular concentric striæ except on the posterior side, which is covered with large radiating ribs, three or four of which, running to the basal angle, are large and more prominent.

Figure. Natural size.

Localities. From the coal mine of Pariatambo, the Cerro of the Ventanillo, and near Ollon.

CRASSATELLA, Lam.

C. CAUDATA, n. s., Pl. 40, fig. 18.

Shell elongate cuneiform, beaks very anterior with the posterior cardinal margin sloping nearly straight to the posterior end, which is narrow and obliquely truncate; anterior end sloping nearly straight from the beaks to near the base, where it is most prominent and rounded. A rounded angle runs from the beaks to the posterior basal angle, behind which the surface is flat or very slightly concave. Surface marked only by coarse lines of growth. Lunule deep and cordate, its margin rounded. The young shells are proportionally much shorter and broader and not so acute posteriorly as the adults.

Figures. Natural size.

Localities. From the black shales of the coal mine of Pariatambo and from the limestones of the Cerro del Ventanillo.

Remarks. A common species, easily recognizable by its elongate, wedge-shaped form. It is perhaps the most abundant fossil in the black shales of the coal mine, associated with *Ammonites carbonarius*. I have never seen the hinge, but have not hesitated in referring it to the above genus, on account of its marked external characters.

C., sp. indet.

A very ventricose cast, apparently of this genus, comes from the Cretaceous locality of the hacienda of Macanga. It retains none of the shell, but from its shape and from the impression of the hinge teeth, it seems to be a *Crassatella*.

CARDITA, Brug., Lam.

C. EXOTICA, d'Orb. sp., Pl. 41, f. 1, 2.
Astarte id., d'Orb., Am. Mer., p. 83, Pl. 18, f. 11-12.
    id., d'Orb., Foss. Col., p. 48, Pl. 3, f. 11-12.

Shell obliquely elongate subquadrate; beaks about a fourth of the distance from the anterior end; directly under the beaks the anterior end is concave; most prominent a little above the middle, below which point it curves regularly into the base; posterior cardinal margin sloping convexly; posterior end broad and obliquely truncate; surface marked by about twenty-four square ribs, roughened by the crossing of irregular ridges of growth; of these ribs, six or seven, smaller than the others, are on the posterior face of the shell; between the large ribs are broad, concave interspaces; lunule small, deep, cordate; inner margin crenulate.

Figures. Natural size.

Localities. Hill near the Hacienda del Imperial near Cañete, and near the town of Coniaca, Dept. of Huancavelica. The latter at the height of 3200 metres. D'Orbigny's specimen came from las Palmas, Prov. of Socorro, U. States of Colombia, and was only an internal cast.

Remarks. Dr. Raimondi refers the first to the Cretaceous, and the second to the Jurassic. Generically this shell belongs to that group, or subgenus, separated by Blainville under the subgeneric title of *Cardiocardites* (not *id.* Meek), and to which Dr. Gray subsequently gave the name of *Agaria.* The Carditas have been so divided up into genera and subgenera that, as in this case, the distinction rests almost, if not entirely, on the outline of the shell, and, while the typical form is marked enough, a regularly gradated series of species, recent or fossil, can be produced to show that this is absolutely of no value.

CARDITA (CYCLOCARDIA) CIRCULARIS, n. s., Pl. 41, fig. 3, 3a.

Shell nearly circular, compressed; beaks somewhat anterior, not very prominent; anterior and posterior ends and base forming continuously three-fourths of a nearly perfect circle; cardinal margin sloping convexly; directly under the beaks the anterior end is slightly excavated; lunule very small; surface covered by twenty-seven or twenty-eight regular radiating ribs, each producing a corresponding crenulation on the inner margin.

Figures. Natural size.

Locality. "Snow mountains to the left of the road between Chonta and Quevopalco; Province of Huamalies; at a height of more than 5000 metres." Referred doubtfully to the Cretaceous.

Remarks. Its remarkably circular form will serve to distinguish this shell from all of its congeners. The specimen has a polished appearance, as if it had been long carried in the pocket, and the fine details of the surface are completely worn off.

TRIGONIA, Brug.

T. BRONNII, Agas., Mém. sur les Trigonies, p. 18, Pl. 5, fig. 19.
Lyrodon clavellatum, Bronn, Lethæa, Pl. 20, f. 3.
id., Goldf., Petr. Germ., p. 200, Pl. 136, f. 6 *a. b.*
Not *T. clavellata*, Sby., Min. Conch., Pl. 87.

A beautiful specimen, agreeing in every detail of outline and ornament with Agassiz's figure and description. Agassiz says the species is peculiar to the Upper Jurassic. Our specimen is from the "immediate vicinity of Jauja; height of 3500 metres."

T. LORENTI, Dana, Wilkes' Exped. Rep., p. 721, Pl. 15, f. 2.

From the Island of San Lorenzo, near Callao. From its type, evidently Jurassic. It is closely allied to *T. sinuata*, Park., but differs in having the ribs bent abruptly upwards posteriorly towards the area, and in having them placed closer together on the middle of the shell and more sinuous than in Parkinson's species.

T. sp. indet.

There is a small lot of fossils in a bad state of preservation, marked "neighborhood of Tingo; Prov. of Huari; Dept. of Huaraz; alt. 3500 metres; Jurassic?". Among them is a small *Trigonia*, recognizable principally by the moulds of the hinge. Not enough of the outline is preserved to ascertain the shape, and the impressions of the surface only suffice to show that it was covered with strong transverse ribs. Fortunately the associated fossils are in a more recognizable state, there being among them *Ammonites Ventanillensis*, nob., *A. carbonarius*, nob., *Tellina Peruana*, nob., and *Crassatella caudata*, nob., all of the present paper, and characteristic of the Liassic beds of the coal mine of Pariatambo and the limestones of the Cerro del Ventanillo. The present rock is a black carbonaceous shale very similar to that of the first of the two mentioned localities, but differs in being a little less calcareous, and in the presence of pyrites which lines the cavities left by the decomposition of several of the shells.

T., sp. indet.

A second species, marked as from the immediate neighborhood of Ollon, and probably Cretaceous, is too imperfect for more than a very doubtful identification. From the number of ribs and general form, I believe it to be the *T. Tocaimana*, Lea, Tr. Amer. Philos. Soc. 1840, p. 256, Pl. 9, f. 8, and which may possibly include *T. Delafossei*, Boyle and Coquand; Mem. Geol. Soc. France, 2 ser., v. 4, Pl. 8, f. 27. This last is represented as differing a little in the costation from Lea's figure; but that was so imperfect a specimen that the matter must be left an open question until better material shall be available.

MYOPHORIA, Bronn.

M. spiralis, n. s., Pl. 41, fig. 4, 4a.

Shell triangular, slightly oblique; beaks in advance of the middle, spirally incurved in advance of and under the umbones; anterior end narrowly and regularly rounded; base sloping slightly upwards behind, where it is nearly straight; most prominent in front; posterior end obliquely truncated; posterior side concave, the concavity bounded by a rounded-angular ridge which runs from the umbone to the posterior-basal angle; surface covered with large, pretty regular lines of growth.

Figures. Natural size.

Locality. From the coal mine of Pariatambo. Liassic.

Remarks. The finest bivalve from this rich and interesting locality. The genus has been usually considered as characteristic of the Trias, all of the species in d'Orbigny's Prodrome being placed in his two stages—5 and 6. But here we have it associated with *Crassatella* and with *Ammonites*, one species of which is only doubtfully separated from a known Jurassic form, and with a *Gyrodes*, true somewhat aberrant, but nevertheless belonging to a genus heretofore not known below the Cretaceous.

PTERIA, Scop.
Avicula, Brug.

P. INCONSPICUA, n. s., Pl. 41, fig. 5.

Shell small, flattened, very oblique; beaks terminal; anterior side sloping so that the most prominent part of the base is directly under the angle of the wing; posterior side broadly emarginate above and rounded below. Surface marked by fine lines of growth and a few large faint radiations on the most convex part.

Figure. Twice natural size.
Locality. With the preceding.

BARBATIA, Gray.

B.? RAIMONDII, n. s., Pl. 41, fig. 6.

Shell small, narrow, compressed; beaks one-fourth of the length from the anterior end; umbones low and broad; a broad shallow depression runs from the umbones downwards and backwards, reaching the base about a third of the length of the shell from the anterior end; anterior end produced and submucronate above, sloping inwards below with a broad curve; base broadly and shallowly emarginate just behind a point opposite the umbones, at the termination of the superficial depression; behind this it is broadly but not prominently convex; posterior end rounded, except adjoining the hinge line, where it is slightly emarginate; area long and very narrow; surface marked by numerous fine radiating ribs, a little broken by lines of growth.

Figure. Twice natural size.
Locality. From the coal mine of Pariatambo.

Remarks. Externally this shell resembles d'Orbigny's figure of *Cucullaea Tocaymensis* (Amer. Merid., Pl. 21, figs. 1–3), but differs in having a narrower area, less prominent beaks, and in being of an entirely different form posteriorly.

TRIGONARCA, Con.

T. ORBIGNYANA, n. s., Pl. 41, fig. 7, 7a, 8, 8a.

Shell very large, triangular; beaks placed about a third of the length from the anterior end, not approximating; anterior end nearly vertical, most prominent near the base; posterior face abruptly truncated; base nearly straight, slightly emarginate and a little the most prominent directly under the beaks; area small, short; internal plate moderate; internal margin entire. Surface unknown; the cast shows traces of numerous rather small radiating ribs.

Figures. Natural size.

Locality. Neighborhood of Ollon. Considered by Dr. Raimondi "Jurassic?", but more probably Cretaceous, both from this and the associated fossils.

Remarks. A large, not very perfect cast, the upper part of the posterior end being broken away, so as to destroy the shape of the point of junction between the area and the posterior margin. Associated with it is a cast of a smaller shell (figs. 8, 8a), an inch across, apparently of the same species, but with a more convex base and flatter sides; the latter difference, however, being one that might be anticipated, from the difference of age. The species cannot be confounded with *Cuc. dilatata*, d'Orb., from the Cretaceous of Bogota, its nearest South American ally, having much higher and more approximating beaks, a straighter base, and less produced posterior end. In *Cucullæa dilatata* the beaks are very wide apart and the lower posterior end much more produced.

T. BREVIS, D'Orb. (sp.), Amer. Merid., p. 89, Pl. 20, f. 2-4.

There is a small shell, one inch in diameter, from the "high table land two leagues from Cajamarca, at a height of more than 3500 metres," which resembles d'Orbigny's figure so closely that I believe it to be the young of that species. D'Orbigny and Raimondi both consider it Cretaceous, and the fossils associated with it seem to confirm the opinion.

T. PERUANA, n. s., Pl. 41, fig. 9, 9a.

Shell trapezoidal, very oblique; beaks anterior, approximated, and incurved; area a little longer than half the length of the shell; posterior end very sloping; base nearly straight. Surface unknown.

Figures. Natural size.

Locality. Neighborhood of Ollon; considered Cretaceous by Dr. Raimondi.

Remarks. A single cast, retaining none of the surface, but showing by its form and by the posterior plates, the generic characters sufficient to place it with certainty, and by its outlines enough specific characters to enable it to be identified. No similar species has been described from the South American secondary rocks.

?ARCA OVALIS, n. s., Pl. 41, fig. 10, 10a.

Shell moderate in size, convex, oblique, very much elongated from beak to base;

base most prominent directly under the posterior end of the hinge line. Area short, narrow, and, at least in part, marked in the usual manner of the Arcas. Shell substance thin; surface marked by concentric lines of growth, crossed by some fine obsolete radiations anteriorly; inner margin entire; a very small plate borders the inner edge of the posterior muscular scar.

Figures. Natural size.

Locality. From the hacienda of Macanga, and from the vicinity of Ollon, from a rock which Dr. Raimondi considers Cretaceous.

Remarks. In referring this shell provisionally to *Arca* I have done so in the full conviction that it will be placed in an, as yet, undescribed genus as soon as more material shall have been obtained. It has the form of an oblique *Limopsis*, but I have been able to see about the anterior third of the area, which shows two or three of those incised lines which are characteristic of the Arcas and Cucullæns. The middle of the area not being visible, I cannot assert that it has not also the *Limopsis* fosset, or at least a rudiment of it. The shell is unusually thin for the family. It wants the marginal crenulation generally seen, and it possesses a rudiment of the internal plate of *Cucullæa*, in the shape of a little ridge bordering only the anterior margin of the posterior muscular scar, and succeeded, in the usual position of the *Cucullæa* plate, by a very faint broadly rounded thickening which runs up into the beak. It will be seen that we have here enough material to characterize this as a new genus, all except the hinge teeth, and the details of the area, the latter of which could, however, be pretty safely inferred.

Another little Arcoid, of which I have only seen one imperfect valve, occurs in the black shale of Pariatambo.

NUCULA, Lam.

N. PERUANA, n. s., Pl. 41, fig. 1.

Shell very small, cuneiform; beaks placed a third of the length from the anterior end, which is rounded; posterior end narrow and rounded; base nearly straight; posterior cardinal margin sloping; surface marked by fine lines of growth.

Locality. Near Tingo; Prov. of Huari; Dept. of Huaraz. Associated with the undetermined *Trigonia* and various species characteristic of the Liassic rocks of Pariatambo; also a single specimen from the latter locality.

Remarks. A cast, showing the outline and characteristic hinge, from the former locality, and another specimen retaining the surface, from the latter. From being almost exactly of the same size, and having about the same degree of convexity as *Corbula nuculoides*, with which it is associated, this shell may be easily confounded with it; but they can be at once distinguished by the more terminal position of the beaks of this, and its cuneiform shape—while in that, the beaks are nearly central, and the posterior cardinal margin, instead of sloping down a trifle convexly, giving a general wedge shape to the shell in this, is in that markedly concave.

N., *sp. indet.*

A large heavy convex species, with very prominent hinge teeth and subcentral beaks, represented by a single cast from near Ollon.

LIMA, Brug.

From the "limestone of Socabon of the mines of the hill of San Antonio of Morococho, on the road from Lima to Jauja, at a height of 4200 metres," there is a fragment, the impression of perhaps half of the surface of a *Lima* or *Plagiostoma* of the type characteristic of the European Lias. It was, when entire, a little more than two inches long, and marked by numerous fine, equal, radiating ribs.

Also, from the Liassic limestone of the Cerro de la Ventanilla is another *Lima* an inch long. The shell is preserved, but it is attached by its outer surface and only shows the interior of the valve, and without the ears. The surface, as seen through the shell, is marked by rather large radiating ribs, strongest on the posterior half of the shell.

PECTEN, Linn.

P. RAIMONDII, n. s., Pl. 42, fig. 1, 1a.

Shell oblong, flattened, equilateral, slightly inequivalve, closed; surface of both valves covered with strong radiating ribs, about 15 or 16 on each valve. These ribs are surmounted by three ridges, and the intervening grooves carry also one or three small radiating lines; all crossed by strong lines of growth becoming slightly squamose on crossing the ridges; anterior right auricle emarginate.

Figure 1, natural size; 1a, a magnified cross section of the ribs.

Locality. From the "hill of Potosi, on the road from Lima to Jauja, in the mineral region of Morococho."

Remarks. The inequal convexity of the two valves places this shell in the subgenus *Chlamys*, Bolt., according to H. and A. Adams. It is of the type of *P. Faujasii* and *P. cretosus*, which it resembles in the style of the ribs, while it differs in their details and in general form, being a shorter and rounder shell.

P. ARGENTARIUS, n. s., Pl. 41, fig. 12, 12a.

Shell flattened, base rounded; sides and ears unknown; surface marked by about 15 or 16 large ribs; angular on top and sloping on the sides; interspaces angular at the bottom; on the sides of the ribs are faint traces of two radiating lines; all crossed by minute, subsquamose lines of growth.

Diameter about 1.3 inch.

Locality. From the "hill of San Antonio, with silver mines, Morococho; altitude 4500 metres," associated with *Rhynchonella Antonii*. No geological age given.

Remarks. I have ventured to name this shell although the outline is almost entirely destroyed, since it cannot fail to be recognized by its slightly convex shell and the peculiar angular ribs and interspaces, a cross section of which forms a zigzag line.

NEITHEA, Drouet.

N. QUINQUECOSTATA, Sowerby, sp.

Pecten id., Sby., Min. Conch., Pl. 56.
Janira id., d'Orb., Pal. Fr., Ters. Cret., p. 632, Pl. 444, f. 1-5.
Neithea id, Gabb, Synopsis Cret. Moll., 1861, p. 148.

A single small specimen, about half grown, retaining both valves. It agrees fully with the European specimens, in all the details of form and ribs. Each large rib has a smaller one on its lateral slope, and the space between each pair of large ribs is occupied by two of slightly smaller size. The specific distinctions in this genus rest on comparatively trivial characters, in the details of surface ornament, and I should have hesitated in referring this shell to Sowerby's species, had I not compared it with authentic specimens from England and France.

Locality. Elevated table-land two leagues from Cajamarca. Cretaceous.

Remarks. We have here an excellent key for establishing a geological horizon, in the presence of a well-known upper greensand species; although it does not follow by any means that the two deposits were absolutely synchronous in their age of deposition. The species may have originated in either of the two regions, and have emigrated, becoming extinct in one before it made its appearance in the other. That this does occur with animals as well as plants is a well-recognized fact; nevertheless, the presence of a species whose geological horizon is established, gives us the means of fixing approximately the age of any new deposit in which it may be discovered.

N. ALATA, Von Buch (sp.).

Pecten alatus, Von Buch, Petr. rec. en Amer., p. 3, Pl. 1, f. 1-4.
P. Dufrenoyi, d'Orb., Am. Mer., p. 106, Pl. 22, f. 5-7.
P. alatus, Bayle & Coq., Mem. Soc. Geol. Fr., vol. 4, p. 14, Pl. 5, f. 1-2.
P. Dufrenoyi, Hup.; Gay's Hist. de Chili, p. 291.
Janira alata, Remond; Pal. de Chili (pamph.), p. 18.
Neithea alata, Gabb, Synopsis Cret. Moll., p. 147.

Three specimens, one from "between Molinas and Paucara;" a second from "a place named Turaino, two leagues below Iscuchaca;" and the third "from the hill of Santa Barbara, where there is a mine of mercury," all in the Dept. of Huancavelica. To the last no geological age is assigned, but Dr. Raimondi refers the first two doubtfully to the Jurassic. Bayle and Coquand referred the shell to the Lias, while d'Orbigny in the Prodrome places it in the Neocomien, in which opinion I followed him in my "Synopsis." Huppé also considers it Lias.

N., sp. indet.

A rolled specimen of the deep valve of an undescribed species, characterized by about 25 nearly uniform large ribs and no apparent intermediate sculpture. It is nearest to *N. Texana*, Roem. (sp.), but that has but 15–17 ribs. It is from the "town of Bagua, on the left bank of the river Utcubamba, Dept. of Chachapoyas."* The shell is too imperfect to describe, but is important as fixing the Cretaceous age of the rocks from which it is derived.

PLICATULA, Lam.

P. TORTA, n. s., Pl. 42, fig. 5.

Shell long, subtriangular, slightly oblique, inequivalve; lower valve slightly convex, upper valve flat; right side nearly straight, left side and base broadly rounded; surface marked by large, coarse squamose lines of growth, crossed by small radiating ribs, which are interrupted by the concentric lines and present a fimbriated appearance.

Figure. Natural size.

Locality. Quebrada of Colpamayo, near Chota, Dept. of Cajamarca. Cretaceous.

ANOMIA, Linn.

A. PERUANA, n. s., Pl. 42, fig. 6.

Shell variable in shape; circular to irregularly elliptical; beaks low, marked, and submarginal; surface marked by irregular lines of growth and concentric undulations.

Measurements. Average diameter, about 0.8 in. One specimen, one inch long by 0.75 wide.

Locality. "Quebrada del Alfalfar, a quarter of a league south of Chachapoyas," on the surface of a distorted cast of a large shell, apparently *Cucullæa Orbignyana*. Cretaceous.

Remarks. The surface of the cast is covered by a group of these little shells,

---

* So says the label, while the catalogue accompanying places it in the "Department of the Amazons."

which had evidently attached themselves to the inner face of the larger one after the death of its occupant. The lower valves are lost in all but one example, and there only traces of it remain.

PLACUNANOMIA, Brod.

P. (PARANOMIA?) LIMA, n. s., Pl. 42, fig. 7.

Shell flat, equivalve, subtriangular; upper margins irregularly sloping; base broadly rounded; surface covered with rough subsquamose lines of growth, crossed by small radiating ribs which, at short intervals, rise into little spines like the teeth of a rasp, or better, like the low spines of some species of *Spondylus*.

Measurements. Length of largest specimen, 2.0 in.; width, 1.7 in.

Localities. From the neighborhood of Ollon; the Hacienda of Macango, and Quebraba of Colpamayo, near Chota, Dept. of Cajamarca. Cretaceous.

Remarks. None of the specimens show the hinge or the internal structure, so that we do not know whether or not it has the internal plate of Conrad's subgenus. It resembles the typical species of that group in the sculpture of the surface, but both valves seem to be equally convex and equally strongly marked with the scabrous ribs.

OSTREA, Linn.

O. CALLACTA, Con., Pl. 42, fig. 2, 2a.
  id., Con., Proc. Acad. Nat. Sciences, 1875, p. 139, Pl. 22, f. 1.
O., sp. indet., Dana, Wilkes' Exped. Rep., Pl. 15, f. 7.

Shell large, oval, nearly equilateral; beaks central, small. Lower valve with one large median ridge, from the lower part of which branches another on each side, the terminations of the three occupying the entire basal margin; a large curved rib runs from the beaks, describing a quarter of a circle, and terminating on the middle of the side; above this on each side are two smaller ribs, rapidly diminishing in size. Each rib on the lower valve corresponds to an interspace on the upper valve, and each interspace of the lower to a rib on the upper. Entire surface covered by rough lines of growth. Area broad, flat; ligament pit very broad and shallow, oblique; margin squamose, not crenulated; muscular impression moderate in size.

Measurements. Diameter from beak to base, 4.5 in.; from side to side, 4.0 in.

Locality. From the Cretaceous at the Hacienda of Macanga, Prov. of Pataz. Mr. Conrad's type came from "the Pampa del Sacramento, Eastern Peru," and he conjectured it to belong to the "Pebas Group" of brackish water Tertiary, first made known by me in 1868.

Remarks. A fine large species, at once recognizable by its outline and its few, large, symmetrically arranged ribs. It was figured without a name in the report of the Wilkes' Exploring Expedition.

O. LARVIFORMIS, n. s., Pl. 42, fig. 3.

Shell small, very inequilateral, oblique, arcuate, nearly equivalve; beaks terminal; posterior side and base forming together a semicircle; anterior side produced, somewhat excavated above; surface marked by about seven large radiating ribs, alternating on the opposite valves; the posterior two or three dividing into two near the margin. All crossed by rough lines of growth. Interior unknown.

Figure. Twice natural size.

Locality. "Hill of three crosses, between Agnamiro and Huallacan, Prov. of Huamalies; height more than 4000 metres." Considered Cretaceous by Dr. Raimondi.

Remarks. A pretty little oyster of the type of *O. larva*, but much shorter, broader, and heavier than that species. From a notch near the beak, it seems to have grown attached to a twig, in the manner of the recent mangrove oysters.

EXOGYRA, Say.

E. POLYGONA, Von Buch, Petr. rec. en Amer., p. 5, Pl. 2, f. 18–19.

Two fine specimens of this strongly characterized species from between Combayo and Polloc; Dept. of Cajamarca. Dr. Raimondi says "calcareous rocks which seem to belong to the Jurassic formation." D'Orbigny refers the species to the Neocomien. Should the small *Exogyra* found with *Ammonites Raimondianus* prove to be this species, it may have to be placed higher in the Cretaceous. See remarks on *E. plicata*.

E., *sp. indet.*, Pl. 8, fig. 4, 4a.

A small shell from the same locality as the preceding, showing the exsert spiral beak characteristic of the genus, but too imperfect to describe. It seems to be somewhat of the type of *E. arietina*, Roem., of the Texas Cretaceous, but not so developed in the beak. The surface is very convex, especially in the middle, and seems to have been smooth. Its entire length is barely three-fourths of an inch.

E. PLICATA, Lam.

For synonymy see Palæontology of California, vol. 2, p. 275.

Two specimens from the same locality as the other two species, and which seem to belong to this well-known form. Their surfaces are somewhat obscured by the matrix, and one at least is marked by unusually small ribs. This, however, is not a valid specific distinction in this genus.

E. PARASITICA, Gabb? Pl. 42, fig. 8, 8a.
E. parasitica, Gabb, Pal. California, vol. 1, p. 205, Pl. 26, f. 192, and Pl. 31, f. 273.

Attached to the large *Cucullæa Orbignyana* is a group of little parasitic Exogyras, all of the same size, and one of which is figured, showing the upper valve and a side view of the lower. In an object possessing so few specific characters I can find none by which to separate this shell from that described by me under the above name from the Cretaceous formation of California.

GRYPHÆA, Lam.

G. RIVOTII, Bayle and Coquand (sp.).
Ostrea, id., B. & C., Mem. Geol. Soc. Fr., 2 s., v. 4, p. 24, Pl. 1, fig. 7, 8.
O. cymbium, B. & C., *pars*. (not Desh.), *loc. cit.*, Pl. 5, f. 6–7.

From the silver mines of Huantajaya. From the figures they give, I strongly suspect that these authors have also included other specimens (Pl. 4, f. 1, 2, 4) in "*O. cymbium*" that belong to this species.

GRYPHÆA, Lam.

G., *sp.?* Pl. 41, fig. 13, 13a.

Five specimens, all lower valves, exactly alike, from the Cretaceous of the Hacienda of Macango, Prov. of Pataz, look like miniature examples of *G. vesicularis*. In all details of shape and proportions they agree so exactly that, were it not for the marked difference in size I should not hesitate in referring them to that species. But their uniformity of size seems to point to a good specific difference. I therefore propose the provisional name of *G. vesiculoides*, in case they may eventually prove distinct.

Figures. Natural size.

TEREBRATULA, Brug.*

T. RAIMONDIANA, n. s., Pl. 42, fig. 9, 9a, 9b.

Shell ovate; beak prominent, overhanging; foramen large, sides most prominent in the middle, below sloping nearly straight to the base, which is subtruncate; small valve with two large plications, which begin about the middle and run to the outer sides of the basal margin; between and on each side of these is a broad, shallow depression, making the edges of the shell sinuous and emarginating the base of the larger valve. On the large valve there is barely a trace of the plications seen on the smaller valve, the surface being almost evenly convex. Surface marked by lines of growth and a few very obscure radiating lines.

---

* I have attributed the genus *Terebratula* to Brugiere on the ground that, although the name was used for nearly a century before, by various pre-Linnæan authors, Brugiere was the first to restrict it to near its present limits. Even Linnæus confounded it with *Anomia*, and the author of the Encyclopedie Méthodique was the first to make a truly scientific use of the name.

Figures. Natural size.
Locality. "A little distance from the town of Ollon."
Remarks. This species is broader and more rounded on the sides than *T. ornithocephala*, Sowerby, the lateral margins of the valves are more sinuous, and it is flatter towards the base, and the plications on the smaller valve are wanting in that species. In the absence of plications on the dorsal valve, it differs from all the other species described from South America.

RHYNCHONELLA, Fisch.

R. ANTONII, n. s., Pl. 42, fig. 10, 10a.

Shell triangular, widest about a third of its length from the base, which is rounded; sides sloping nearly straight from the beak at an angle of about 70° from each other; beak narrow, elevated; area large. Large valve bearing a broad shallow sinus, ending in a correspondingly broad, short tongue which encroaches on the small valve. Smaller or ventral valve regularly convex on the surface except close to the base, where the side of one rib is carried down to fit the side of the tongue of the opposite valve. Surface ornamented by about eighteen radiating ribs, of which usually six are in the groove of the dorsal valve and seven corresponding with those on the ventral; in one example there are eight on the dorsal and nine on the ventral valves in the middle.

Figures. Slightly magnified.
Localities. Three specimens marked "Cerro de San Antonio, with mines of silver." This locality is also given for *Pecten argentarius*, but the rock is different lithologically. Another specimen is marked "Quebrada de Colpamayo, near Chota," the locality of *Placunanomia lima*. This agrees lithologically with the other specimens of the same species, but also differs from the other fossils of the locality. Still another specimen in a coarse red sandstone is marked "Cerro de Vivuco, a league and a half from Ollon." Thus two of the three lots are said to come from Cretaceous localities or neighborhoods, but all three differ in the character of the rock from the other fossils. We have, therefore, no reliable clue to the geological age of the species.

ECHINODERMATA.

ECHINUS, Linn.

E. BOLIVARII, d'Orb., Amer. Merid., p. 88, Pl. 21, fig. 11-13.

From the "Hacienda of Uchupata; Prov. of Huari. Cretaceous."

E.? *sp. indet.*

Another shell, resembling this genus, but with all the finer details destroyed by crystals of carbonate of lime, which has completely replaced the shell, leaving only the external form and the coarser details of the ambulacra. It is from the Liassic locality of the Cerro del Ventanillo.

BOTRIOPYGUS, d'Orb.

B. ELEVATUS, n. s., Pl. 43, fig. 1, 1a.

Shell very rounded subpentagonal, slightly narrowed in front and produced in the middle behind; broadly convex above, concave below; details of ornament of upper surface unknown; lower surface closely studded with small tubercles.

Figures. Natural size.

Locality. From the Cretaceous table, two leagues from Cajamarca.

B. COMPRESSUS, n. s., Pl. 43, fig. 2, 2a.

Shell broadly elliptical, ends equal; upper surface very broadly convex; under, slightly concave; details of surface ornament unknown.

Figures. Natural size.

Locality. Near Calca, a few leagues south of Chachapoyas. Cretaceous.

Remarks. This species is larger than the preceding, and differs in being flatter, broader, and regularly elliptical. In that, the anterior end is narrower than the posterior, and the region of the anal opening is prominent; whereas in this, both ends are of the same shape and width.

DISCOIDEA, Gray.

D. NUMISMALIS, n. s., Pl. 43, fig. 3, 3a, 3b.

Shell faintly pentagular, very depressed, apex slightly elevated, surface sloping straight, or a little concavely and then convexly to the margin; under surface concave; anal opening large, elongate. Surface marked by numerous small, regularly placed tubercles, scattered uniformly; larger and more scattered on the under side.

Figures. Natural size.

Locality. "Cattle estate of Yauca, at the foot of the perpetual snow, one league and a half from the town of Queropalca, Prov. of Huamalies; height of 4000 metres." Marked "Cretaceous?"

Associated with this is a smaller specimen of the same genus, but apparently another species. The surface has been rubbed smooth and the centre bored so as to enable it to be used as a bead. It is marked 1000 metres higher, same vicinity.

Two others, agreeing with it in character occur, but both are so imperfect on the surface that I have not ventured to describe them. The first is from the elevated table-land, two leagues from Cajamarca; the other from the town of Pion, Prov. of the Amazons. They are certainly Cretaceous.

ENNALASTER, d'Orb.

E. PERUANUS, n. s., Pl. 43, fig. 4, 4a, 4b, 4c.

Shell rounded subpentagonal; anterior end rounded, emarginate; posterior truncate; sides most prominent in advance of the middle; top convex, highest posteriorly; mouth small, rounded; anus rounded-triangular, with the apex below. Upper surface with tubercles sparsely scattered, some smaller ones bordering the anterior groove; below, a triangular space between the mouth and posterior end is closely studded with tubercles, while the remainder of the surface carries but few.

Figures. Natural size.

Locality. From the Cretaceous table, two leagues from Cajamarca, and near Ollon.

Remarks. Closely allied to *E. Texanus*, Roem. sp. (*Toxaster*), but is less markedly pentagonal, more narrowed behind, and differs in the shape of both the apertures.

PERIASTER, d'Orb.

P. AUSTRALIS, n. s., Pl. 43, fig. 5, a, b, c.

Shell ovate-pentagonal, narrowed and truncated behind; rounded and emarginate in front; ambulacra deeply impressed, the posterior pair very much shorter than the others; the inter-ambulacral spaces angulated in the middle near their apices; upper surface high posteriorly, sloping downwards in advance; mouth subreniform, transverse; anus elliptical, greatest diameter vertical. Surface closely covered with small tubercles; on the under side is a triangular space with its apex towards the mouth, bordered by a broad naked strip.

Figures. Natural size.

Localities. From the hills of the District of Huancaspata, Prov. of Pataz, and from the Hacienda of Uchupata, Prov. of Huari; also, the Hacienda of Macanga, the neighborhood of the town of Huallanca, Prov. of Huamalies and near the town of Bagua, Dept. of Chachapoyas. Cretaceous.

Remarks. This species differs from the Texan Cretaceous *P. Texanus*, Roem. sp. (*Hemiaster*) in having a less pentagonal outline, being broader and much higher posteriorly, with the apex placed further back, and in having the anal groove less pronounced. It also differs somewhat in the shape and position of both apertures.

PART III. CARBONIFEROUS FOSSILS.

The following little lot is, with the exception of a single species, from the "Island of Titicaca in the lake of the same name, at an altitude of 4000 metres." The geological age is fixed at once by the presence of the universally distributed *Fusulina cylindrica*. The deposit consists of a reddish-brown and gray limestone, and the genus *Productus* is well represented both in number of species and specimens.

TEREBRATULA, Brug.

T. TITICACENSIS, n. s., Pl. 42, fig. 11, 11a.

Shell small, convex, subpentagonal; beaks high, sides concavely sloping; area small; surface covered only with fine lines of growth. Adult shell, large valve with a deep median sinus, ending in a long, tongue-like projection of the base, and bounded by two large rounded ridges; corresponding with this sinus, there is one equally large ridge in the small valve. The presence of this sinus and ridge cause the lateral margins of the valves to be strongly sinuous. In the young shell the margins are straight and the sinus is wanting.

Figures. Natural size.

Remarks. Allied to *T. Peruvianus*, d'Orb., but larger and more regularly slender. That species is referred by the author to the Devonian.

RHYNCHONELLA, Fisch.

R. ANDII, d'Orb. (sp.).
Terebratula, id., d'Orb., Amer. Merid. 45, Pl. 3, f. 14–15.
Atrypa, id., d'Orb., Prod. Pal. Strat., v. 1, f. 147.

PRODUCTUS, Sby.

P. CANCRINI, Verneuil, Geol. Russ., p. 273,, Pl. 16, f. 8; Pl. 18, f. 7.
P. *spinosus*, Kutorga, Vert. der Min. Ges. St. Pet., 1842, p. 18, Pl. 5, f. 2, *not id.* Sby.
P. *Villiersii*, d'Orb., Am. Mer., p. 53, Pl. 4, f. 12-13.
P. DOLIVIENSIS, d'Orb., Amer. Merid., p. 52, Pl. 4, f. 5–9.

I refer a single rather imperfect specimen to this species, since it has all of the characters assigned by the author so far as can be made out.

P. PAPILIO, n. s., Pl. 42, fig. 12, 12a.

Shell subquadrate; beak small; cardinal line produced into two expanded submucronate points; sides emarginate, narrowing below, base deeply emarginate; surface of large valve with a deep median groove and two shallower radiating depressions running to the sides; the basal two-fifths of the surface is bent down at a right angle; entire surface marked by small closely-placed ribs, between which are small punctures; these ribs are crossed by small concentric undulations, taking the direction of the lines of growth. Smaller valve unknown.

Figure. Natural size.

Remarks. Allied to *P. Boliviensis* and *P. Andii*, d'Orb., in the expanded hinge line, but differs from both in the beak being very small, instead of being large and incurved. A marked character, by which it can be distinguished from all the other known species of South America, is the sudden manner in which the basal portion of the surface is bent down at a right angle.

P. RETICULATUS, n. s., Pl. 42, fig. 13, 13a.

Shell large, subquadrate; hinge line not longer than the greatest width of the shell; sides and base broadly rounded; base but very faintly emarginated. Lower valve with a shallow median groove, bent down to nearly a right angle, but not abruptly; beak small; upper valve nearly flat, slightly concave towards the margin and slightly grooved in the middle of the basal portion. Surface marked by numerous small ribs, crossed by concentric ridges of about the same size, producing a cancellated appearance.

Figures. Natural size.

Remarks. Allied to *P. Inca*, d'Orb., but less coarsely striate, more angulated in cross section of the lower valve, from beak to base; the beak is very much smaller, and it wants entirely the traces of spines on the surface seen in that species.

P., *sp. indet.*

A small species, barely half an inch wide, too imperfect for description, from the "Heights of Cullachaca, three leagues from Huanta."

POLYZOA.
RETEPORA.

R. FLEXUOSA, d'Orb., Amer. Merid., p. 57, Pl. 6, fig. 6–8.

CRINOIDEA.

Two or three masses containing stems of crinoids, one of which is four inches long, half an inch in diameter, and cylindrical.

FORAMENIFERA.
FUSULINA, Fisch.

F. CYLINDRICA, Fisch., Oryct. Gouv. Mosc., p. 126, Pl. 18, f. 1–5.

PART IV. BIBLIOGRAPHY OF S. AMERICAN PALÆONTOLOGY.

1839. Von Buch; Pétrifications recueillies in Amérique, par Mr. Alexandre de Humboldt et par Mr. Charles Degenhardt. Secondary fossils.
1840. I. Lea; Transactions American Philos. Society, Philadelphia, 2 ser., vol. 7, p. 251.
Notice of the Oolitic Formation in America, with Descriptions of some of its Organic Remains.
Fossils from near Bogotá, Cretaceous (and Jurassic?).
1842. D'Orbigny, Alcide; Voyage dans l'Amérique Meridionale, vol 4. Palæontology, Tertiary to Palæozoic.
1842. D'Orbigny, A. Coquilles et Echinodermes fossils de Colombie, recueillies par M. Doussingault. In great part a republication of the preceding, with much of the same text and plates.
1844. Forbes; Quart. Journal Geological Society, London, vol. 1, p. 174.
Report on the Fossils from Santa Fé, Bogotá, presented to the Geological Society by Evan Hopkins, Esq. Cretaceous.
1846. Darwin; Geological Observations on South America, 4th part of Report of Voyage of the Beagle.
Tertiary fossils by G. B. Sowerby; Secondary by E. Forbes.
1846. Conrad; Proceedings Acad. Nat. Sciences, Philada.; vol. 3, p. 19.
Descriptions of New Species of Fossil and Recent Shells, etc. Two species; Tertiary.
1847. D'Orbigny, A.; Voyage au Pole Sud, etc., l'Astrolabe et la Zélée. Cretaceous Fossils.
1850. Von Buch; Zeitschrift der Deutschen Geologischen Gesellschaft, vol. 2, 339. "Die Anden in Venezuela." *Ammonites Tucuyensis.*
1851. Bayle & Coquand; Memoirs Geological Society of France, 2d ser., vol. 4, p. 1. Mémoire sur les Fossiles Secondaires recueillies dans le Chile, par M. Ignace Domeyko. Cretaceous and Jurassic.
1854. Huppé in Gay; Historia Fisica y Politica de Chile. Secondary and Tertiary.
1855. Conrad, T. A.; U. S. Naval Astronomical Expedition, under Lieut. Gillis, vol. 2, p. 282. Secondary and Tertiary Fossil from Chile.
1856. Karsten; Verhandlungen der Versammlung deutscher Naturforscher in Wien. "Die geognostischen Verhältnisse Neu-Grenada's." Cretaceous Mollusca and Foramenifera.
1860. Gabb; Proceedings Academy of Natural Sciences, Phila., 1860, p. 197.
Descriptions of some new species of Cretaceous Fossils from South America in the collection of the Academy. Chile.
1860. Philippi, R. A.; Viaje al Desierto de Atacama; also, German and perhaps other editions. Secondary Fossils.
1861. Burmeister & Giebel; Die Versteinerungen von Juntas im Thal des Rio Copiapo.
1867. Rémond; Paleontologia de Chile, pamphlet, catalogue.
1868. Gabb; American Journal of Conchology, vol. 4, p. 157.
Descriptions of Fossils from the Clay Deposits of the Upper Amazon. Late Tertiary.
1869. Gabb; *loc. cit.*, vol. 5, p. 25.
Descriptions of New Species of South American Fossils. No. 1. Tertiary from Peru. (Preliminary to the present memoir.)

1870. Nelson; Transactions of the Connecticut Academy, vol. 2, p. 186.
  On the Molluscan Fauna of the Later Tertiary of Peru.
1870. Hyatt, Alph.; Hartt's Geology and Physical Geography of Brazil, p. 385.
  Report on the Cretaceous Fossils from Maroïm, Prov. of Sergipe, Brazil, in the collection of Prof. Hartt.
1870. Conrad; American Journal Conchology, vol. vi., p. 192.
  Descriptions of new Fossil Shells from the Upper Amazon. Tertiary.
1871. Woodward; Annals and Mag. N. History, iv. ser., vol. vii., pp. 59 and 101.
  The Tertiary Shells of the Amazon Valley.
1872. Dall; American Journal of Conchology, vol. vii., p. 89.
  Notes on the Genus Anisothyris, Conrad, with a description of a new species.
1874. Conrad; Proceedings Academy of Natural Sciences, Philada., vol. 26, p. 25.
  Remarks on the Tertiary Clays of the Upper Amazon, with Descriptions of New Shells.
1875. Rathbun; Proc. Boston Soc. of Natural History, vol. 17, p. 241.
  Preliminary Report on the Cretaceous Lamellibranchiates collected in the vicinity of Pernambuco, Brazil.
1875. Hyatt; *loc. cit.*, p. 365.
  The Jurassic and Cretaceous Ammonites collected in South America by Prof. Jas. Orton; with an appendix upon the Cretaceous Ammonites of Prof. Hartt's collection.
1875. Hartt and Rathbun; Annals of the New York Lyceum of Natural History.
  On the Devonian Trilobites and Mollusca of Ererè, Province of Pará, Brazil.
1875. Conrad; Proceedings Academy of Natural Sciences, Philada.
  Description of a New Fossil Shell from Peru. *Ostrea callacta.*
Date? Dana; Report of the United States Exploring Expedition; Geology. Secondary. (There is no date on the title-page of this report, but it must have been published some time about 1850.)

PART V.—SYNOPSIS OF SOUTH AMERICAN PALÆONTOLOGY.

PALÆOZOIC.

ARTICULATA.

DALMANIA PAITUNA, Hartt and Rath., Ann. N. Y. Lyc. 1875, p. 111.	Devonian; Brazil.
CALYMENE VERNEUILII, d'Orb., Am. Mer., p. 31, Pl. 1, f. 4–5.	Silurian; Bolivia.
C. MACROPHTHALMA, Brogn., Hist. des Crust, Pl. 15, f. 1–5.	
Id. d'Orb., Am. Mer., p. 31, Pl. 1, f. 6–7.	Sil.; Bolivia.
HOMALONOTUS OIARA, Hartt and Rath., An. N. Y. Lyc., 1875, p. 114.	Dev.; Braz.
ASAPHUS BOLIVIENSIS, d'Orb., Am. Mer., p. 32, Pl. 1, f. 8–9.	Sil.; Bol.
CRUZIANA, d'Orb. (*Bilobites*, Cord., not DeKay).	
C. RUGOSA, d'Orb., Am. Mer., p. 30, Pl. 1, f. 1.	
Bilobites on plate.	Sil.; Bol.
C. FURCIFERA, d'Orb., Am. Mer., p. 30, Pl. 1, f. 2–3.	
Bilobites on plate.	Sil.; Bol.

MOLLUSCA.

CEPHALOPODA.

TENTACULITES ELDREDGIANUS, H. and R., Ann. N. Y. Lyc., 1875, p. 126.	Dev.; Brazil.
BELLEROPHON MORGANIANUS, H. and R., Ann. N. Y. Lyc., 1875, p. 117.	Dev.; Brazil.
B. COUTINHOANUS, H. and R., Ann. N. Y. Lyc., 1875, p. 117.	Dev.; Brazil.
B. GILLETIANUS, H. and R., Am. N. Y. Lyc., 1875, p. 118.	Dev.; Brazil.

GASTEROPODA.

NATICA BUCCINOIDES, d'Orb., Am. Mer., p. 43, Pl. 3, f. 8–9.	Carb.; Bol.
N. ANTISIENSIS, d'Orb., *loc. cit.*, p. 43, Pl. 3, f. 10.	Carb.; Bol.

Neither of the above species belong to the genus to which the author referred them, nor to any other living genus. The information given us in the work is too imperfect to enable one to decide satisfactorily where they should be placed. If the restored outline of the first is correct, it does not belong even to this family.

SOLARIUM ANTIQUUM, d'Orb., Am. Mer., p. 42, Pl. 3, f. 1–3.
S. PERVERSUM, d'Orb., *loc. cit.*, p. 43, Pl. 3, f. 5–7.

Both of these are from the Carboniferous of Bolivia and seem to be antique forms, possibly identical with, or at least closely allied to, the modern genus *Architectonica*, from which they differ in the absence of the characteristic sculpture.

PLEUROTOMARIA ANGULOSA, d'Orb., *loc. cit.*, p. 43, Pl. 3, f. 4.	Carb.; Bol.
P. ROCHANA, H. and R., Ann. N. Y. Lyc., 1875, p. 115.	Dev.; Braz.
PLATYCERAS SYMMETRICUM, Hall, 15 Ann. Rep. N. York State Cabinet, 1862, p. 34.	
Id. H. and R., Ann. N. Y. Lyc., 1875, p. 116.	Dev.; Brazil.

LAMELLIBRANCHIATA.

GRAMMYSIA (*Pholadella?*) PARALLELA, Hall.	
Id. H. & R., A. N. Y. Lyc., 1875, p. 120.	Dev.; Brazil.
EDMONDIA PONDIANA, H. & R., A. N. Y. Lyc., 1875, p. 121.	Dev.; Brazil.
E. SYLVANA, H. & R., A. N. Y. Lyc., 1875, p. 122.	Dev.; Brazil.
MEGALODON ANTIQUA, d'Orb., Prod., v. 1, p. 130.	
Trigonia antiqua, d'Orb., Am. Mer., p. 44, Pl. 3, f. 12–13.	Carb.; Bol.
MODIOMORPHA PIMENTANA, H. & R., A. N. Y. Lyc., 1875, p. 123.	Dev.; Brazil.
PALÆONEILO SULCATA, H. & R., *loc. cit.*, p. 124.	Dev.; Brazil.
P.? SIMPLEX, H. & R., *loc. cit.*, p. 125.	Dev.; Brazil.
NUCULITES NYSSA, Hall; H. & R., *loc. cit.*, p. 119.	Dev.; Brazil.
N. ERERENSIS, H. & R., *loc. cit.*, p. 120.	Dev.; Brazil.
PECTEN PAREDESI, d'Orb., Am. Mer., p. 44, Pl. 3, f. 11.	Carb.; Bol.

BRACHIOPODA.

TEREBRATULA TITICACAENSIS, Gabb, n. s.	Carb.; Peru.
SPIRIFER BOLIVIENSIS, d'Orb., Am. Mer., p. 37, Pl. 2, f. 8–9.	Dev.; Bol.

MADE BY DR. ANTONIO RAIMONDI IN PERU. 307

S. INCRASSATUS, Eichw., Russ., p. 276, Pl. 4, f. 12.
S. PENTLANDI, d'Orb., Am. Mer., p. 48, Pl. 5, f. 15.   Carb.; Bol.
S. QUICHUA, d'Orb., *loc. cit.*, p. 37, Pl. 2, f. 21.   Dev.; Bol.
S. ROISSYI, Lev., Mem. G. Soc. Fr., 1835, p. 39, Pl. 2, f. 18.
Terebratula *Peruviana*, d'Orb., Am. Mer., Pl. 3, f. 17, 19.
S. *Roissyi*, d'Orb., *loc. cit.*, p. 46.   Carb.; Bol.
S. STRIATUS, Sby., M. Con., p. 125, Pl. 270.
S. *condor*, d'Orb., A. Mer., p. 46, Pl. 5, f. 11–14.   Carb.; Bol.
ATRYPA ANDII, d'Orb., Prod. Pal., v. 1, p. 147.
Terebratula *id.*, d'Orb., Am. Mer., p. 45, Pl. 3, f. 14–15.   Carb.; Bol., Peru.
A. GAUDRYI, d'Orb., Prod. v. 1, p. 147.
Terebratula *id.*, d'Orb., A. Mer., p. 45, Pl. 3, f. 16.   Carb.; Bol.
A. ANTISIENSIS, d'Orb., Prod., v. 1, p. 95.
Terebratula *id.*, d'Orb. A. Mer., p. 36, Pl. 2, f. 26–28.   Dev.; Bol.
A. PERUVIANA, d'Orb., Prod., v. 1, p. 95.
Terebratula *id.*, d'Orb., A. Mer., p. 36, Pl. 2, f. 22–25.   Dev.; Bol.
ORTHIS HUMBOLDTII, d'Orb., A. Mer., p. 27, Pl. 11, f. 16–20.
Spirifer, on plate.   Sil.; Bol.
O. CORA, d'Orb., A. Mer., p. 48, Pl. 3, f. 21–23.   Carb.; Bol.
O. BUCHII, d'Orb., *loc. cit.*, p. 49.   Carb.; Bol.
O. INCA, d'Orb., *loc. cit.*, p. 38, Pl. 2, f. 10–12.   Dev.; Bol.
O. PECTINATA, d'Orb., *loc. cit.*, p. 39, Pl. 2, f. 13–15.   Dev., Bol.
O. LATECOSTATA, d'Orb., *loc. cit.*, p. 39.   Dev.; Bol.
PRODUCTUS ANDII, d'Orb., *loc. cit.*, p. 54, Pl. 5, f. 1–3.   Carb.; Bol.
P. BOLIVIENSIS, d'Orb., *loc. cit.*, p. 52, Pl. 4, f. 5–9.
P. *Boliviensis & Gaudryi*, on plate.   Carb., Bol. & Peru.
P. CANCRINI, Vern., Geol. Russ., p. 273, Pl. 16, f. 8, and Pl. 18, f. 7.
P. *spinosus*, Kutonga, Verh. der Min. Ges., St. Pet. (1842), p. 18, Pl. 5, f. 2.
Not P. *spinosus*, Sby.
P. *Villiersii*, d'Orb., A. Mer., p. 53, Pl. 4, f. 12–13.
P. *Cancrini*, Gabb, present paper.   Carb.; Bol., Peru.
P. CAPACII, d'Orb., A. Mer., p. 50, Pl. 3, f. 24–26.   Carb.; Bol.
P. CORA, d'Orb., *loc. cit.*, p. 55, Pl. 5, f. 8–9.   Carb.; Bol.
P. HUMBOLDTII, d'Orb., *loc. cit.*, p. 54, Pl. 5, f. 4–7.   Carb.; Bol.
P. INCA, d'Orb., *loc. cit.*, p. 51, Pl. 4, f. 1–3.   Carb.; Bol.
P. PAPILIO, Gabb, new sp.   Carb.; Peru.
P. PERUVIANUS, d'Orb., Am. Mer., p. 52, Pl. 4, f. 4.   Carb.; Bol.
P. RETICULATUS, Gabb, new sp.   Carb.; Peru.
LEPTÆNA VARIOLATA, d'Orb., Am. Mer., p. 49, Pl. 4, f. 10–11.
Productus id., d'Orb., on plate.
LINGULA SUBMARGINATA, d'Orb., Prod. Pal., v. 1, p. 14.
L. *marginata*, d'Orb., Am. Mer., p. 28, Pl. 2, f. 5.   Sil.; Bol.
L. MUNSTERI, d'Orb., *loc. cit.*, p. 29, Pl. 2, f. 6.   Sil.; Bol.
L. DUBIA, d'Orb., *loc. cit.*, p. 29, Pl. 2, f. 7.   Sil.; Bol.

DESCRIPTION OF A COLLECTION OF FOSSILS,

POLYZOA AS RADIATA.

RETEPORA FLEXUOSA, d'Orb., Am. Mer., p. 57, Pl. 6, f. 6-8.
   Id.   G., present paper.   Carb.; Bol., Peru.

RADIATA.

CYATHAXONIA STRIATA, d'Orb., Prod. Pal., v. 1, 158.
Turbinolia striata, d'Orb., Am. Mer., p. 56, Pl. 6, f. 1-5.   Carb.; Bol.
CERIPORA RAMOSA, d'Orb., *loc. cit.*, p. 56, Pl. 6, f. 9-10.   Carb.; Bol.
GRAPTOLITHUS FOLIACEUS, Murch., Sil. Syst., Pl. 26, f. 3.
G. Murchisonii, Beck, Murch., Sil. Syst., Pl. 26, f. 4.
G. dentatus, d'Orb., Am. Mer., p. 32, Pl. 2, f. 1.   Sil.; Bol.

FORAMENIFERA.

FUSULINA CYLINDRICA, Fisch., Oryct. Gouv. Mosc., p. 126, Pl. 18, f. 1-9.
   Id.   G., present paper.   Carb.; Peru.

MESOZOIC.

CRUSTACEA.
CIRRIPEDIA.

Pinna minuta, Gabb, Proc. Phila. A. N. S., 1860, p. 198, Pl. 3, f. 10, from the Cretaceous of Chile, is apparently the carinal plate of a *Scalpellum*, and since no species of this genus has been described from this region, it will have to retain the specific name and be called *S. minutum*. This is an unfortunate name, since the species is of average size, but is another illustration of the bad results which arise from even well-intentioned character names, and not less so of the regrets with which more mature students sometimes have to look back on their juvenile efforts.

MOLLUSCA.
CEPHALAPODA.

BELEMNITES CHILENSIS, Con., U. S. Naval Astron. Exped., p. 284.
   Id.   Philippi, Viajo al Desierto Atacama, p. 143, Pl. 1, f. 4.   Jurassic? Chile.
B. GIGANTEUS, Schlot., 1803; Taschenbuch, p. 284.
   Id.   Huppé, Gay's Chile, p. 25.   Ool.; Chile.
For synonomy see d'Orb., Pal. Fr., Terr. Jus., p. 112.
HELICERUS FUEGIENSIS, Dana, Wilkes' Exped., p. 720, Pl. 15, f. 1.   Cret.; Pat.
NAUTILUS CHILENSIS, Huppé, Gay's Chile, p. 30.   Jur.; Chile.
N. DEKAYI, Morton, Syn. Cret., p. 33, Pl. 8, f. 4.
N. perlatus, Mort., *id.*, p. 33, 13, f. 4.
N. Orbignyanus, Forbes, Darwin's S. A., p. 265, Pl. 5, f. 1.
N. lævigatus, d'Orb., Voy. Astrol. and Zélée, Pl. 6, f. 1-2.
N. Valenciennii, Hup., Gay's Chile, p. 28, Pl. 1, f. 1.   Cret.; Chile
N. INDICUS, d'Orb., Prod., vol. 2, p. 211.
N. Sowerbyanus, d'Orb., Voy. Ast. and Zel., Pl. 4, f. 1-2.   Cret.; Chile.
N. SEMISTRIATUS, d'Orb., Pal. Fr., Ter. Jur., Pl. 26, f. 1-3.

N. Domeykus, d'Orb., A. Mer., p. 103, Pl. 22, f. 1-2.
N. semistriatus, Hup., Gay's Chile, p. 29.
N. semistriatus, Bayle & Coquand, M. G. Soc. Fr., 2 s., v. 5, p. 9, Pl. 1, fig. 4.   Jurassic; Chile.
N. STRIATUS, Sby., M. Con., p. 183, Pl. 182.
   Id. B. & Coq., M. G. Fr., 2 s., v. 5, p. 8, Pl. 2, f. 6.
   Id. Hup., Gay's Chile, p. 30.   Lias.; Chile.
N. TENUI-PLANATUS, Dana, Wilkes' Exped., p. 721, Pl. 15, f. 4.   Jur.; Peru.
BACULITES ANCEPS, Lam., A. S. V., v. 7, p. 648.
   Id. d'Orb., Astr. and Zel., Pl. 4, f. 8-12.
   Id. Hup., Gay, p. 41.   Cret.; Chile, Col.
B. Granatensis, Karst., Geog. Verh. N. Gren., p. 105, Pl. 2, f. 1.
B. Maldonadi, K., *loc. cit.*, p. 105, Pl. 2, f. 2.
B. LYELLI, d'Orb., Ast. & Zel., Pl. 4, f. 3-7.
B. vagina, Fbs., Darwin's S. A., Pl. 5, f. 3.
Not id., Fbs., Tr. G. Soc. Lond.   Cret.; Chile.
B. VAGINA, Fbs., Tr. G. Soc. Lond., v. 7, p. 114, Pl. 10, f. 4.
Not id., Fbs., Darwin's Geol. Obs. S. A.
B. ornata, d'Orb. Astr. & Zel., Pl. 6, f. 3-6.   Cret.; Chile.
PTYCHOCERAS HUMBOLDTIANUS, Karst., Geog. Verh. N. Gren., p. 101, Pl. 1, f. 1.   Cret.; Col.
HAMULINA DEGENHARDTII, (V. Buch sp.), d'Orb., Prod., v. 2, p. 102.
Hamites id., Von Buch, Petr. Rec. par Humb., p. 17, figs. 23-25.
   Id. Forbes, Quart. Jour. Geol. Soc., v. 1, p. 175.   Cret.; Colombia.
   Id. Karst., Geog. Verh. N. Gren., p. 102.
HAMITES HUMBOLDTIANA, Lea sp.
Orthoceras id., Lea, Tr. A. P. S., 2 s., v. 7, p. 253, Pl. 8, f. 1.
Hamites d'Orbignyana, Forbes, Quart. J. G. Soc., v. 1, p. 175.

On comparing Mr. Lea's original specimen with Forbes' paper, it becomes clear that his determination of the *Ancyloceras* as being the same as Lea's species is incorrect. The *Ancyloceras* has tubercles on the septate portion of the shell, and on the body chamber the ribs are irregularly placed. This part is also not perfectly straight. Forbes' description and figure of his *Hamites* correspond exactly with Mr. Lea's fragment, now before me. I have therefore revised his determination.

CRIOCERAS DUVALII, Lev., Mem. G. Soc. Fr., v. 2, p. 313, Pl. 22, f. 1.
   Id. Bayle & Coq., M. G. S. F., 2 s., v. 4, p. 34, Pl. 3, f. 1-4.
   Id. Hup., Gay's Chile, p. 40.   Cret.; Chile, Col.
   Id. var. *undulata*, Karst., Geog. Verh. N. Gren., p. 102, Pl. 1, f. 3.
ANCYLOCERAS HUMBOLDTIANA, Fbs., Q. J. G. S., v. 1, 175.
Not *Orthoceras id.*, Lea, Tr. Amer. Phil. Soc., 2 s., v. 7, p. 253, Pl. 8, f. 1.
See *Hamites Humboldtiana*.   Cret.; Colombia.
A. BEYRICHII, Karst., G. V. N. Gren., p. 103, Pl. 1, f. 4.   Cret.; Col.
A. SIMPLEX, d'Orb., Pal. Fr., Terr. Cr., p. 503, Pl. 125, f. 5-8.
Hamites elatior, Fbs., Darwin's S. A., p. 265.   Cret.; Patagonia.
LINDIGIA HELICOCEROIDES, Karst., Geog. Verh. N. Gren., p. 103, Pl. 1, f. 5.   Cret.; Col.

310   DESCRIPTION OF A COLLECTION OF FOSSILS,

AMMONITES AALENSIS, Ziet., Wurt., Pl. 28, f. 3.
   Id.   d'Orb., P. Fr., T. Jur., p. 238, Pl. 63.
A. candidus, d'Orb., on plate.
   Id.   Burmeist. & Gieb, 29.   Jur.; Chile.
A. ACOSTÆ, Karst., Geog. Verh., N. Gren., p. 111, Pl. 5, f. 1.   Cret.; Col.
A. ACUTISSIMUS, Gabb, new sp.   Cret.?; Peru.
A. ÆOOCEROS, Phil., Atacama, p. 142, Pl. 2, f. 3.
   Id.   Gabb, present paper.   Jur.; Chile and Peru.
A. ÆQUATORIALIS, Von Buch, Petr., p. 15, f. 11–12.   Cret.; Col.
A. ANCEPS, Rein. (sp.), d'Orb., Pal. Fr., T. Jur., p. 462, Pl. 166, 167.
Nautilus id., Reineke, Naut. & Am., p. 82, Pl. 7, f. 61.
Perisphinctes id., Hyatt, P. Bost. N. H. Soc., v. 17, p. 368.   Jur.; Brazil.
A. ANDII, Gabb, new sp.   Jur.; Peru.
A. ANNULATUS, Sby., M. Con., vol. 3, p. 41, Pl. 222.
A. annularis, Phil., Atacama, p. 141.   Col.; Chile.
A. ATACAMENSIS, Phil., Atacama, p. 142, Pl. 1, f. 1–2.
A. sp. indet., Dana, W. Exped., Pl. 15, f. 6.   Jur.; Chile & Peru.
A. ATTENUATUS, Hyatt (sp.)   Peru.
Buchiceras id., Hyatt, P. B. Soc. N. H., v. 17, p. 372.   Cret.; Brazil.
A. BARBACOENSIS, Karst., Geog. Verh. N. Gren., p. 111, Pl. 4, fig. 5.   Cret.; Col.
A. BIFURCATUS, Schlot., in Zieten, Pl. 3, fig. 3.
A. Garantianus, d'Orb., P. Fr., T. Jur., p. 377, Pl. 121.
A. bifurcatus, Bayle & Coquand, M. G. S. Fr., 2 s., v. 4, p. 20, Pl. 2, f. 2.
   Id.   Hup., Gay's Chile, p. 38.   Ool.; Chile.
A. BILOBATUS, Hyatt (sp.)
Buchiceras id., H., Proc. Bost. N. H. Soc., v. 17, p. 370.   Cret.; Brazil, Peru.
A. BISULCATUS, Brug., Enc. Meth., v. 1, p. 39, No. 13.
   Id.   Hup., Gay's Chile, p. 32.   Jurassic; Chile.
A. BOGOTENSIS, Fbs., Q. J. G. Soc., v. 1, p. 178.   Cret.; Colombia.
A. BOUSSINGAULTII, d'Orb., Am. Mer., p. 68, Pl. 1, f. 1–2.
   Id.   d'Orb., Foss. Col., p. 32, Pl. 1, f. 1–2.   Cret ; Col.
A. BRACKENRIDGII, Sby., M. Con , p. 187, Pl. 184.
   Id.   Phil., Atacama, p. 141.   Ool.; Chile.
A. BRODIEI, Sby., M. Con., Pl. 351.
   Id.   Phil., Atacama, p. 140.   Ool ; Chile.
A. BUCHIANA, Fbs., Q. J. G. Soc., v. 1, p. 177.   Cret.; Colombia.
A. CANALICULATUS, Munst., in Zieten, p. 37, Pl. 28, f. 6.
   Id.   Hup. Gay's Chile, p. 38.
A. opalinus, Pusch., Quenst., Petr., Pl. 7, f. 10.
   Id.   B. & Coq., M. G. S. Fr., 2 s., v. 4, p. 10, Pl. 2, f. 1.   Lias.; Chile.
A. CAQUEZENSIS, Karst., Geog. Verh. N. Gren., p. 104, Pl. 1, f. 7.
A. ubaquensis, K., loc. cit., p. 104, Pl. 1, f. 8.   Cret.; Col.
A. CARBONARIUS, Gabb, new sp.   Lias.; Peru.

A. ceras, Giebel, Fauna der Vorw., Ceph., p. 757.
Arnioceras id., Hyatt, Proc. B. S. N. H., v. 17, p. 366. Lias.; Peru.
A. codazzianus, Karst., Geog. Verh. N. Gren., p. 108, Pl 3, f. 4, 5. Cret.; Col.
A. colombianus, d'Orb., Am. Mer., p. 17, Pl. 17, f. 12–14.
  Id.   d'Orb., Foss. Coll., p. 41, Pl. 2, f. 12–14. Cret.; Colombia.
A. communis, Sby., M. Con., Pl. 117, f. 2–3.
  Id.   Phil., Atacama, p. 141. Lias.; Chile.
A. corniferus, Gabb, new sp. Jur.; Peru.
A. damsianus, d'Orb., A. Mer., p. 69, Pl. 2, f. 1–2.
  Id.   d'Orb, Foss. Col., p. 33, Pl. 2, f. 1–2. Cret.; Col.
A. domeykanus, B. & C., M. G. S. Fr., 2 s., v. 4, p. 10, Pl. 2, f. 3–5.
  Id.   Hup., Gay's Chile, p. 36. Lias.; Chile.
A. dupinianus, d'Orb., P. Fr., T. Cr., p. 276, Pl. 81, f. 6–8.
  Id.   Karst., G. Verh. N. Gren., p. 412, Pl. 5, f. 5. Cret.; Col.
A. fimbriatus, Sby., M. Con., v. 2, p. 145, Pl. 164.
  Id.   Hup., Gay's Chile, p. 23. Jur.; Chile.
A. galeatus, V. Buch, Petr. rec. par Humb., p. 12, Pl. 2, f. 20.
  Id.   d'Orb., Am. Mer., p. 75, Pl. 17, f. 3–7.
  Id.   d'Orb., Foss. Col., p. 37, Pl. 2, f. 3–7.
A. *Tocaymensis*, Lea, Tr. A. Phil. Soc., 2 s., v. 7, p. 253, Pl. 8, f. 2. Cret.; Colombia.
A. *Americanus*, Lea, *loc. cit.*, p. 255, Pl. 8, f. 6.
A. *compressissimus*, d'Orb., Pal. Fr., T. Cr., v. 1, p. 210, Pl. 61, f. 4–5.
A. *Didayanus*, d'Orb., *loc. cit.*, p. 360, Pl. 108, f. 4–5.
A. *Leai*, Forbes, Quart. Jour. Geol. Soc., v. 1, p. 178.
A. *galeatoides*, Karst. Verh. N. Gren., p. 107, Pl. 3, f. 1.
A. *caicedi*, Karst., *loc. cit.*, p. 107, Pl. 3, f. 2.
A. *Lindigii*, Karst., *loc. cit.*, p. 108, Pl. 3, f. 3. Cret.; Col.
A. gayi, Gabb.
A. *tripartitus*, Huppé, Gay's Chile, p. 35, Pl. 1, f. 2.
Not A. *tripartitus*, Raspail. Jur.; Chile.

  This resembles *A. Boussingaultii*, d'Orb. (Cretaceous), but the two authors place their species in different formations. Apart from this, which may be an error on the part of one or the other, the species differ, according to the figures, in this species having fewer nodes on the middle of the whorl, and in each node sending a rib to the suture. The figure of d'Orbigny's species makes the umbilical half plain. Nevertheless, I have seen such great variation in the surface ornaments of the same species of Ammonites that, had the two forms been referred to the same geological age, I should have preferred calling the two by one name.

A. gemmatus, Hup., Gay's Chile, p. 35, Pl. 1, f. 3. Jur.; Chile.
A. gibbonianus, Lea, Tr. A. P. S., 2 s., v. 7, p. 254, Pl. 8, f. 3.
  Id.   Hyatt, Hartt's Brazil, p. 389.
  Id.   Gabb, present paper. Cret.; Col., Peru, Brazil.
  The North American form referred to this by Marcou and subsequent authors is a distinct but allied species. It increases much more rapidly in the size of the whorls, and they have a

312   DESCRIPTION OF A COLLECTION OF FOSSILS,

flatter cross section. Strange to say, although it has been quoted half a dozen times, nobody has had more than imperfect fragments, and the septum is, as yet, unknown. Perhaps the impression named by Dana *A. Pickeringii* (*q. v.*) should be placed here.

A. HALLI, Meek & Hayden, Proc. Phila. Acad., 1856, p. 70.
Phylloceras? id., Meek, Rep. Hayden's Survey, p. 458, Pl. 24, f. 3, *a, b, c*.
    Id.   Hyatt, Hartt's Brazil, p. 388.   Cret.; Brazil.
A. HARTTI, Hyatt (sp.).
Ceratites id., Hyatt, Hartt's Brazil, p. 386.   Cret.; Brazil.
A. HOPKINSII, Fbs., Q. J. G. S., v. 1, p. 176.   Cret.; Colombia.
A. HUGARDIANUS, d'Orb., P. Fr., T. Cret., v. 1, p. 291, Pl. 86, f. 1.
    Id.   V. Buch, Zeits. D. Geol. Gesell., v. 2, p. 342.   Cret.; Venez.
A. HYATTI, Gabb, new sp.   Jurassic; Peru.
A. INFLATUS, Sby., M. Con., p. 178.
    Id.   V. Buch, Zeits. D. Geol. Gesellschaft, v. 2, p. 341.   Cret.; Venez.
A. LEONHARDINUS, Karst., Geog. Verh., N. Gren., p. 106, Pl. 2, f. 5.   Cret.; Col.
A. LIGATUS, d'Orb., Pal. Fr., Terr. Cret., p. 126, Pl. 38.
A. Inca, Fbs., Q. J. G. S., v. 1, p. 176.   Cret.; Colombia.
A. LOSCOMNI, Hyatt (sp.).
Phylloceras id., II., Proc. Bost. N. H. Soc., v. 17, p. 368.   Lias.; Peru.
A. MACROCEPHALUS, Schlot., Taschenb., p. 70.
    Id.   Hup., Gay's Chile, p. 36.
A. corrugatus, Hup., on plate.
Stephanoceras id., Hyatt, P. B. S. N. H., v. 17, p. 368.
A. macrocephalus, Gabb, present paper.   Ool.; Chile, Bolivia, Peru.
A. MAYORIANUS, d'Orb., P. F., T. Cr., v. 1, p. 267, Pl. 79.
    Id.   V. Buch, Zeits. D. Geol. Gesell., v. 2, p. 342.   Cret.; Venez.
A. MISERABILIS, Quenst., Jura, p. 71.
Arnioceras id., Hyatt, P. B. S. N. H., v. 17, p. 367.   Lias.; Bolivia.
A. NOEGGERATIIII, Karst., Geog. Verh., N. G., p. 104, Pl. 1, f. 6.   Cret.; Col.
A. MOSQUERÆ, Karst., Geog. Verh. N. G., p. 111, Pl. 4, f. 4.   Cret.; Col.
A. OCCIDENTALIS, Lea, Tr. A. P. S, 2 s., v. 7, p. 254, Pl. 8, f. 4.
A. Vanuxemensis, Lea, *loc. cit.*, p. 254, Pl. 8, f. 5.
A. Alexandrinus, d'Orb., A. Mer., p. 75, Pl. 17, f. 8–11.   Cret.; Col.
A. OLLONENSIS, Gabb, new sp.   Cret.? Peru.
A. ORBIGNYI, Gabb, 1861, Proc. A. Phil. Soc., Syn. Cret., p. 14.
A. alternatus, d'Orb., A. Mer., p. 71, Pl. 1, f. 5–6.
Not id., Woodward, nor Portl.   Cret.; Colombia.
A. ORTONI, Hyatt (sp.).
Caloceras id., Hyatt, Pr. B. Soc. N. H., v. 17, p. 367.   Jur.? Peru.
A. PERARMATUS, Sby., M. Con., Pl. 352.
    Id.   Philippi, Atacama, p. 141.   Jur.; Chile.
A. PERUVIANUS, V. Buch, Petr., p. 5, f. 5–7.   Cret.; Colombia.

A. Pickeringii, Dana, Wilkes' Exped., p. 721, Pl. 15, f. 5.   Cret.? Peru.
See note to *A. Gibbonianus.*
A. Pedernalis, V. Buch, Uber. Cerat., p. 31, Pl. 6, f. 8–10.
Ceratites Pierdenalis, Hyatt, Hartt's Brazil, p. 388.
Buchiceras Pierdernalis, Hyatt, P. Bost. N. H. Soc., v. 17, p. 369.
A. *pleurisepta,* Conrad, Emory's Mex. Boundary Rep., p. 159, Pl. 15, f. 1.   Cret.; Brazil.
A. planidorsatus, d'Orb., Am. Mer., p. 72, Pl. 1, f. 6–9.
    Id.   d'Orb., Foss. Col., p. 36, Pl. 1, f. 6–9.   Cret.; Col.
A. plicatilis, Sby., Min. Conch., vol. 2, p. 148, Pl. 166.
A. *biplex,* Sby., *loc. cit.,* p. 168, Pl. 253.
    Id.   Hup., Gay's Chile, p. 34.   Jur.; Chile.
A. pustulifer, B. & C., M. G. S. Fr., 2 s., v. 4, p. 141, Pl. 1, f. 3.   ? Lias.; Chile.
A. radians, Schlot., Petr., p. 28, No. 34.
    Id.   Huppé, Gay's Chile, p. 34.
    Id.   Philippi, Atacama, p. 141.   Lias.; Chile.
A. Raimondianus, Gabb, new sp.   Jur.; Peru.
A. Rhotomagensis, Brogn., Env. Paris, p. 83, Pl. 4, f. 2.
? *Id.*   V. Buch, Petr., p. 7, f. 15.
A. *verrucosus,* Hup., Gay's Chile, p. 39, Pl. 1, f. 4.
Not id. d'Orb., Pal. Fr., T. Cret., p. 191, Pl. 58, f. 1–3.   Cret.; Chile, Colombia.
Ancyloceras Buchianus, d'Orb., Prod. Pal., vol. 2, p. 101.
A. Roissyanus, d'Orb., Pal. Fr., Terr. Cret., v. 1, p. 302, Pl. 89.
    Id.   V. Buch, Zeits. D. Geol. Gesell., v. 2, p. 342.   Cret.; Venez.
A. Roseanus, Karst., Geog. Verh., N. Gren., p. 106, Pl. 2, f. 4.   Cret.; Col.
A. rotundus, Sby., M. Con., Pl. 293, f. 3.
    Id.   Phil., Atacama, p. 141.   Jur.; Chile.
A. Sanctafecinus, d'Orb., A. Mer., p. 70, Pl. 1, f. 3–4.
    Id.   d'Orb., Foss. Col., p. 34, Pl. 1, f. 3–4.   Cret.; Chile.
A. serratus, Hyatt (sp.).
Buchiceras id., Hyatt, P. Bost. S. N. H., v. 17, p. 370.   Cret.; Brazil.
A. Syriaciformis, Hyatt (sp.).
Buchiceras id., H., Pr. B. S. N. H , v. 17, p. 371.   Cret.; Brazil.
A. Toroanus, Karst., G. Verh. N. Gren., p. 109, Pl. 4, f. 2.   Cret.; Col.
A. Treffryanus, Karst., *loc. cit.,* p. 109, Pl. 4, f. 1.   Cret.; Col.
A. Trionæ, Karst., Geog. Verh. N. G., p. 105, Pl. 2, f. 3.   Cret.; Col.
A. Tucumensis, V. Buch, Zeitsch. Deutsch. Geol. Gesell., vol. 2, p. 342, Pl. 10.   Cret.; Venez.
A. variadilis, d'Orb., Pal. Fr., T. Jur., v. 1, p. 350, Pl. 113.
    Id.   Burmeister & Giebel, p. 29.   Lias.; Chile.
A. varicosus, Sby., M. Con., p. 451.
    Id.   V. Buch, Zeits. D. Geol. Gesell., v. 2, p. 341.   Cret.; Venez.
A. Ventanillensis, Gabb, new sp.   Jurassic; Peru.
Trigonellites lanceolatus, Gabb.
Aptychus, sp. indet., Phil., Atacama, p. 143, Pl. 1, f. 3.   Jurassic; Chile.

314   DESCRIPTION OF A COLLECTION OF FOSSILS,

GASTEROPODA.

NEPUNEA CHILENSIS, d'Orb. (sp.).
Fusus id., d'Orb., Voy. Astr. & Zel., Pl. 4, f. 29.   Cret.; Chile.
N. (*Tritonofusus*) DIFFICILIS, d'Orb. (sp.).
Fusus id., d'Orb., Am. Mer., p. 118, Pl. 12, f. 11-12.
  Id.   d'Orb., Ast. & Zel., Pl. 4, f. 27, 28.
  Id.   Hup., Gay's Chile, p. 171.
Neptunea id., Gabb, Syn. Cret., 1861, p. 62.   Cret.; Chile.
PERISSOLAX HOMBRONIANA, d'Orb. (sp.).
Pyrula id., d'Orb., Voy. Ast. & Zel., Pl. 4, f. 31.
P. dilatata, Hup., Gay's Chile, p. 179, Pl. 2, f. 2.
Perissolax Hombroniana, Gabb, Syn. Cret., p. 67.   Cret.; Chile.
P. DURVILLEI, d'Orb. (sp.).
Fusus id., d'Orb., Astr. & Zel., Pl. 5, f. 1.
Perissolax id., Gabb, Syn. Cret., p. 67.   Cret.; Chile.
P. LONGIROSTRIS, d'Orb. (sp.).
Pyrula id., d'Orb., Am. Mer., p. 119, Pl. 12, f. 13.
  Id.   d'Orb., Astr. & Zel., Pl. 4, f. 30.
Perissolax id., G., Syn. Cret., 1861, p. 67.   Cret.; Chile.
P. TROCHOIDES, Gabb, new sp.   Cret.; Peru.
SURCULA ARATA, Gabb.
Pleurotoma id., Gabb, P. A. N. S., Phila., 1861, p. 198, Pl. 3, f. 9.   Cret.; Chile.
DRILLIA SUBEQUALIS, Sby. (sp.).
Pleurotoma id., Sby., Darwin, p. 257, Pl. 4, f. 52.   Cret.; Chile.
BELA ORDIGNYANA, Gabb.
Pleurotoma id., G., loc. cit., p. 198, Pl. 3, f. 7.   Cret.; Chile.
B. ARAUCANA, d'Orb. (sp.).
Pleurotoma id., d'Orb., A. Mer., p. 119, Pl. 14, f. 10-11.
  Id.   d'Orb., Ast. & Zel., Pl. 4, f. 35-36.
  Id.   Hup., Gay, p. 177.   Cret.; Chile.
LUNATIA ARAUCANA, d'Orb. (sp.).
Natica id., d'Orb., Am. Mer., p. 115, Pl. 12, f. 4-5.
  Id.   Hup., Gay's Chile, p. 222.
N. *australis*, d'Orb., Ast. & Zel., Pl. 4, f. 20-21.
  Id.   Gay, p. 223.   Cret.; Chile.
LUNATIA SOLIDA, Sby. (sp.).
Natica id., Sby., Darwin's S. A., p. 255, Pl. 3, f. 40-41.   Cret.; Chile, Patagonia.
PRISCONATICA? GIBBONIANA, Lea (sp.).
Natica id., Lea, Tr. A. P. S. Phil., 2 s., v. 7, p. 256, Pl. 9, f. 10.   Cret.; Colombia.
P. AMPLA, Gabb, new sp.   Cret.; Peru.
P. INCONSPICUA, Gabb, new sp.   Cret.; Peru.
P. OVOIDEA, Gabb, new sp.   Cret.? Peru.
P. PRÆLONGA, Leym. (sp.).

Natica id., Leym., Mem. Geol. Soc. Fr., v. 5, p. 13, Pl. 16, f. 8.
Id. d'Orb., Pal. Fr., T. Cret., v. 2, p. 339, Pl. 172, f. 1.
Id. d'Orb., Am. Mer., p. 78, Pl. 19, f. 1.
Id. Von Buch, Zeits. Deutsch. G. Gesell., v. 2, p. 343.
Id. Hyatt, Hartt's Brazil, p. 385. Cret.; Venez., Col., & Brazil.
GYRODES AUCA, d'Orb. (sp.).
Natica id., d'Orb., Ast. & Zel., Pl. 4, f. 22–23.
N. Chilina, d'Orb., *loc. cit.*, f. 24–26. Cret.; Chile.

I place both these names of d'Orbigny's as synonyms because, although the two sets of figures seem to differ, I do not find the differences sustained by specimens from Chile in the Academy's museum. The cross sections of the body whorl, the only character on which the author relied, seem to vary even more than his figures, and to gradate from one form to the other.

G. CONTRACTA, Gabb, new sp. Lias.; Peru.
G. LIBATA, Gabb, new sp. Cret.; Peru.
RUMA GRANGEANUS, d'Orb. (sp.).
Natica id., d'Orb., Ast. & Zel., Pl. 4, f. 18–19. Cret.; Chile.
? NEVERITA PHASIANELLA, Bayle & Coquand (sp.).
Natica id., B. & C., M. G. S. Fr., 2 s., v. 4, p. 23, Pl. 2, f. 9.
Id. Huppé, Gay's Chile, p. 224. Cret.; Chile.
NEVERITA STRIOLATA, Sby., Darwin's S. A., p. 255, Pl. 5, f. 39. Cret.; Chile.
NATICA OUDIGNYI, Huppé, Gay, p. 224. Cret.; Chile.
SCALA (*Opalia*) AUCA, d'Orb.
Scalaria auca, d'Orb., Ast. & Zel., Pl. 4, f. 16–17. Cret.; Chile.
? S. CHILENSIS, d'Orb., Am. Mer., p. 114, Pl. 14, f. 1–2.
Id. Hup., Gay's Chile, p. 152. Cret.; Chile.
Possibly not a *Scala*, but figure and description are alike too poor to warrant an opinion.
S. (*Clathrus*) PATTONI, Gabb.
S. (*C.*) *Chilensis*, Gabb, P. Phil. Acad. 1860, p. 197, Pl. 3, f. 4.
Not S. Chilensis, d'Orb.
Scala Pattoni, Gabb, Proc. A. P. Soc. Phila., vol. 8, p. 135.
S. Gabbi, Remond, Pal. Chile, p. 31. Cret.; Chile.

In case d'Orbigny's species should prove to belong to another genus, then the old name for this must be restored.

TYLOSTOMA VARIABILIS, Gabb, Pal. Cal., v. 2, p. 261, Pl. 35, f. 6.
Id. Gabb, present paper. Cret.; Peru.
CINULIA CHILENSIS, d'Orb. (sp.).
Avellana id., d'Orb., Ast. & Zel., Pl. 4, f. 32–34. Cret.; Chile.
C. ANTIQUA, Gabb, new sp.
ACTÆONELLA OVIFORMIS, Gabb, new sp. Cret.; Peru.
LOXONEMA POTOSENSIS, d'Orb., Prod. Pal., v. 1, p. 172.
Chemnitzia Potosensis, d'Orb., A. Mer., p. 60, Pl. 6, f. 1–3.
Melania, on plate. - Trias.; Bolivia.
PUGNELLUS UNCATUS, Forbes (sp.), Gabb, Syn. Cret., p. 72.

316   DESCRIPTION OF A COLLECTION OF FOSSILS,

Strombus id., Fbs., Tr. G. Soc. Lond., v. 7, p. 129, Pl. 13, f. 6.
S. semicostatus, d'Orb., Astr. & Zelee, Pl. 5, f. 39-39.
Colombellina uncata, d'Orb., Prod. Pal., v. 2, p. 231.
Pugnellus id., Rémond, Pal. Chile, p. 30.   Cret.; Patagonia.?
   I place this species in the list because Rémond attributes it to Port Famine. At the same time I doubt the propriety, suspecting that a confusion has arisen from the fact that it was refigured by d'Orbigny in the report of the voyage of the Astrolabe and Zelée in a mixture of South American and East Indian species.
PUGNELLUS TUMIDUS, Gabb, Proc. A. N. S. of Phila., 1860, p. 197, Pl. 3, f. 13-14.   Cret.; Chile.
? HIPPOCHRENES BOUSSINGAULTII, d'Orb. sp.
Rostellaria id., d'Orb., Am. Mer., p. 80, Pl. 18, f. 5.
   Id.   d'Orb., Foss. Col., p. 45, Pl. 3, f. 5.   Cret.; Col.
   Not a true *Hippochrenes* of the type of *H. macroptera,* but nearer to that than to any named genus. The expanded lip suddenly narrows in the middle to a narrow tongue. But it will subserve no good end to give generic names to each shape of lip in the protean group of slate shells, as I have already pointed out in my notes on the genus *Anchura.* If species of this type were numerous it would aid classification to give them a distinctive name; but where but a single species is known it results rather in confusion than simplification, to multiply names.
ANCHURA AMERICANA, d'Orb. (sp.).
Rostellaria id., d'Orb., Am. Mer., p. 80, Pl. 18, f. 5.
   Id.   d'Orb., Foss. Col., p. 45, Pl. 3.   Cret.; Colombia.
Not A. (*Rost.*) *Americana,* Evans & Shumard (N. Amer.).
A. ANGULOSA, d'Orb. (sp.).
Rostellaria id., d'Orb., Am. Mer., p. 80, Pl. 18, f. 4.   Cret.; Col.
TURRITELLA PERUANA, Gabb, new sp.   Cret.; Peru.
T. RAIMONDII, Gabb, new sp.   Lias.; Peru.
LITHOTROCHUS HUMBOLDTII, Von Buch (sp.).
Pleurotomaria id., Von Buch, Petr., p. 9, f. 26.
Turritella Andii, d'Orb., Am. Mer., p. 104, Pl 6, f. 11.
Lithotrochus Andii, Con., U. S. Astron. Exped., p. 243, Pl 41, f. 3.
Trochus Struveanus, Zim., Dunk., Pal., p. 185, Pl. 26, f. 2.
Turritella Humboldtii, Boyle & Coq., M. G. S. Fr., 2 s., v. 4, p. 12, Pl. 2, f. 7-9.
   Id.   Gabb, Syn. Cret., p. 90.
T. *Andii,* Hup., Gay's Chile, p. 156.   Cret.; Chile, Peru, Bolivia, and Colombia.
PETROPOMA PERUANA, Gabb, new sp.   Lias.; Peru.
DENTALIUM CHILENSE, d'Orb., Ast. & Zel., Pl. 4, f. 37-38.   Cret.; Chile.
PATELLA AUCA, Gabb, Proc. Acad. N. S. of Phila., 1860, p. 198, Pl. 3, f. 11.   Cret.; Chile.
HELCION CARBONARIUS, Gabb, new sp.   Lias.; Peru.
ACTÆON AFFINIS, Fitt (sp.), d'Orb., Pal. Fr., T. Cr., Pl. 168, f. 46.
Tornatella id, Fitt., Tr. G. S. Lond., v. 4, Pl. 18, f. 9.
Actæon id., d'Orb., A. Mer., p. 79.   Cret.; Colombia.
A. ORNATA, d'Orb., A. Mer., p. 79.   Cret.; Col.
A. SEMINOSA, Gabb.

MADE BY DR. ANTONIO RAIMONDI IN PERU. 317

Eulima id., G., Proc. Acad. N. S. of Phila., 1860, p. 197, Pl. 3, f. 6.   Cret.; Chile.
    With a very *Eulimoid* form, this shell proves to have a strong fold on the inner lip.
BULLA CHILENSIS, d'Orb., Ast. & Zel., Pl. 4, f. 13-15.   Cret.; Chile.

LAMELLIBRANCHIATA.

CULTELLUS AUSTRALIS, Gabb, P. A. N. S., Phila., 1860, p. 198, Pl. 3, f. 8.   Cret.; Chile.
PANOPÆA COQUIMBENSIS, d'Orb., A. Mer., p. 126, Pl. 15, f. 7-8.
    Id.   Huppé, Gay's Chile, p. 373.   Cret.; Chile.
P. SIMPLEX, Hup., Gay, p. 374, Pl. 6, f. 7.   Cret.; Chile.
    Lutraria on plate.
? P. TURGIDA, Hup., Gay, p. 375, Pl. 6, f. 3.
Donax id., on plate.   Jur.; Chile.
P. UNDULATA, Gabb, new sp.   Cret.; Peru.
CORBULA CHILENSIS, d'Orb., Ast. & Zel., Pl. 5, f. 11-12.   Cret.; Chile.
C. COLUMBIANA, d'Orb., Am. Mer., p. 84.   Cret.; Col.
C. CORBULOPSIS, Gabb.
Thracia id., Gabb, Proc. Acad. N. S. of Phila., 1860, p. 198, Pl. 3, f. 1.   Cret.; Chile.
C. NUCULOIDES, Gabb, new sp.   Lias.; Peru.
C. PERUANA, Gabb, new sp.   Lias.; Peru.
C. RAIMONDII, Gabb, new sp.   Lias.; Peru.
ANATINA COLOMBIANA, d'Orb., A. Mer., p. 84, Pl. 18, f. 16-17.
    Id.   d'Orb., Foss. Col., p. 49, Pl. 3, f. 16-17.   Cret.; Col.
PHOLADOMYA ABBREVIATA, Hup., Gay, p. 377, Pl. 6, f. 4.   Jur.; Chile.
P. ACOSTÆ, B. & C., M. G. S. Fr., v. 4, p. 21, Pl. 7, f. 5-6.
    Id.   Hup., Gay's Chile, p. 377.   Jur.; Chile.
P. AUSTRALIS, Gabb, new sp.   Cret.; Peru.
P. FIDICULA, Sby., M. Con., Pl. 225.
    Id.   B. & Coq., M. G. S. Fr., 2 s., v. 5, p. 27, Pl. 7, f. 7.
    Id.   Hup., Gay's Chile, b. 376.
P. attenuata, Hup., Gay's Chile, p. 376, Pl. 6, f. 5.   Jur.; Chile.
P. RAIMONDII, Gabb, new sp.   Cret.; Peru.
P. ZIETENII, Agas. Etud. sur les Myes, p. 54, Pl. 3, f. 13-15.
P. fidicula, Ziet. (not Sby.).
P. Zieteni, Bayle & Coquand, Huppé, Gabb.   Jur.; Peru, Chile.
HOMOMYA INCURVA, Gabb, new sp.   Cret.? Jur.? Peru
H. LÆVIGATA, Huppé (sp.).
Pholadomya id., Hup., Gay, p. 378, Pl. 6, f. 6.   Jur.; Chile.
H. PONDEROSA, Gabb, new sp.   Cret.; Peru.
MACTRA ARAUCANA, d'Orb., A. Mer., p. 125, Pl. 15, f. 3-4.
    Id.   Hup., Gay, p. 349.
    Id.   d'Orb., Ast. & Zel., Pl. 5, f. 2-4.   Cret.; Chile.
    Id.   *var.* Gabb, Proc. Acad. N. S. of Phila., 1860, p. 198.   Cret.; Chile.
M. CHILENSIS, Gabb, *loc. cit.*, p. 198, Pl. 3, f. 5.   Cret.; Chile.

318 DESCRIPTION OF A COLLECTION OF FOSSILS,

Both these species bear a marked resemblance externally to my genus *Cymbophora*, but I have not placed them under that name since the proof, in the hinge characters, has never yet been attained. They are not true *Mactras* in the restricted sense.

? LUTRARIA CUNEIFORMIS, Hup., Gay, p. 351, Pl. 3, f. 3.	Cret.; Chile.
TELLINA LAROILLIERTI, d'Orb., Ast. & Zel., Pl. 5, f. 5–6.	
Nucula id., d'Orb., Am. Mer., p. 128, Pl. 15, f. 9–10.	Cret.; Col.
T. PERNAMBUCENSIS, Rath., Proc. Bost. N. H. Soc., v. 17, p. 256.	Cret.; Brazil.
T. PERUANA, Gabb, new sp.	Lias.; Peru.
? T. VALDIVIANA, d'Orb. (sp.).	
Arcopagia id., d'Orb., Ast. & Zel., Pl. 5, f. 7–9.	Cret.; Chile.
CYPRIMERIA PERUVIANA, Con., Jour. Conch., 1866, p. 105, Pl. 9, f. 1.	Cret.; Peru.
? VENUS ÆREA, Hup., Gay, p. 338.	Cret.; Chile.
? V. AUCA, d'Orb., A. Mer., p. 122, Pl. 12, f. 17–18.	
Id. d'Orb., Ast. & Zel., Pl. 5, f. 9–10.	
Id. Hup., Gay, p. 341.	Cret.; Chile.
? V. CHIA, d'Orb., A. Mer., p. 82, Pl. 18, f. 9–10.	Cret.; Col.
? V. CRETACEA, d'Orb., A. Mer., p. 82.	Cret.; Col.
? V. DUBIA, Hup., Gay, p. 344, Pl. 6, f. 9.	Jur.; Chile.
? V. INSULA, Hup., Gay, p. 343, Pl. 6, f. 10.	Cret.; Chile.
V. CECILIANA, d'Orb. (sp.).	
Mactra id., d'Orb., A. Mer., p. 126, Pl. 15, f. 5–6.	
Venus Orbignyi, Gabb, P. A. N. S., Phila., 1860, p. 199, Pl. 3, f. 2.	Cret.; Chile.
CALLISTA McGRATHIANA, Rath., P. Bost. N. H. Soc., v. 17, p. 255.	Cret.; Brazil.
CARDIUM ACUTICOSTATUM, d'Orb., A. Mer., p. 120, Pl. 12, f. 11–12.	
Id. d'Orb., Ast. & Zel., Pl. 5, f. 17–20.	Cret.; Chile.
C. (*Protocardia*) APPRESSUM, Gabb, new sp.	Lias.; Peru.
C. (*P*) PEREGRINORSUM, d'Orb., A. Mer., p. 81, Pl. 18, f. 6–8.	Cret.; Col., Venez.
Id. V. Buch, Zeits. D. Geol. Gesell., v. 2, p. 343.	
C. (*Trachycardium*) AUCA, d'Orb., A. Mer., p. 121, Pl. 13, f. 14–15.	
C. id., Huppé, Gay, p. 325.	Cret.; Chile.
C. (*T.*) AUSTRALINUM, d'Orb.	
C. *Australinum*, d'Orb., Prod., vol. 2, p. 242.	
C. *Australe*, d'Orb., Ast. & Zel., Pl. 5, f. 21–22.	Cret.; Chile.
C. (*T.*) COLOMBIANUM, d'Orb.	
C. *Colombianum*, d'Orb., A. Mer., p. 82.	Cret.; Col.
C. (*Criocardium*) SOARESANUM, Rath.	
C. id., Rath., P. Bost. N. H. Soc., v. 17, p. 253.	Cret.; Brazil.
C. (*C.*) STRIATELLUM, Philippi.	
C. *striatellum*, Phil., Atacama, p. 143, Pl. 2, f. 6.	Lias.? Chile.
LUCINA AMERICANA, Forbes, Darwin's S. A., p. 266, Pl. 5, f. 24.	Age? Peru.
L. DEMOULINI, d'Orb., Ast. & Zel., Pl. 5, f. 15–16.	Cret.; Chile.
L. EXCENTRICA, Sby., Darwin's S. A., p. 267, Pl. 5, f. 21.	Cret.; Pat.
L. GRANGEI, d'Orb., Ast. & Zel., Pl. 5, f. 13–14.	Cret.; Chile.

MADE BY DR. ANTONIO RAIMONDI IN PERU.

L. PLICATO-COSTATA, d'Orb., A. Mer., p. 83, Pl. 15, f. 13–14.
  Id. d'Orb., F. Col., p. 48, Pl. 3, f. 13–14. Cret.; Col., Venez.
  Id. Von Buch; Zeits. D. G. Gesell., v. 2, p. 344.
L. TENELLA, Rath., Pr. Bost. N. H. Soc., v. 17, p. 253. Cret.; Brazil.
ASTARTE DARWINII, Fbr., Darwin's S. A., p. 266, Pl. 5, f. 22–23. Age? Chile.
A. DUBIA, d'Orb., A. Mer., p. 105, Pl. 6, f. 12–13. Cret.; Chile.
?A. GREGARIA, Phil., Atacama, p. 143, Pl. 2, f. 4. Jur.; Chile.
A. TRUNCATA, V. Buch, Petr., p. 13, f. 17. Cret.; Col.
?CRASSATELLA BOGOTINA, d'Orb. (sp.).
Tellina id., d'Orb., A. Mer., p. 84, Pl. 18, f. 15. Cret.; Col.
C. BUCHIANA, Karst., Geog. Verh. N. Gren., p. 113, Pl. 5, f. 7. Cret.; Col.
  Most probably not a Crassatella.
C. CAUDATA, Gabb, new sp. Lias.; Peru.
? C. VENERIFORMIS, Hup., Gay, p. 362, Pl. 6, f. 11. Cret.; Chile.
CARDITA EXOTICA, d'Orb. (sp.), Gabb, present paper.
Astarte id., d'Orb., A. Mer., p. 83, Pl. 18, f. 11–12. Cret.; Col., Peru.
C. (*Cyclocardia*) CIRCULARIS, Gabb, new sp. Cret.? Peru.
VENERICARDIA MORGANIANA, Rath. (sp.).
Cardita id., Rath., Proc. Bost. S. N. H., v. 17, p. 250. Cret.; Brazil.
V. WILMOTII, Rath. (sp.).
Cardita id., Rath., P. B. S. N. H., v. 17, p. 251. Cret.; Brazil.
MYTILUS ARAUCANUS, d'Orb., Ast. & Zul., Pl. 5, f. 27–28. Cret.; Chile.
LITHOPHAGA AUSTRALIS, Gabb.
Modiola cretacea, G., Proc. Acad. N. S., Phil., 1860, p. 198, Pl 3, f. 3.
Lithodomus Australis, G. Syn. Cret., p. 138. Cret.; Chile.
M. SCALPRUM, Sby., M. Con., Pl. 248, f. 2.
Mytilus id., Goldf., Petr., Pl. 130, f. 9.
  Id. B. & Coq., M. G. S. Fr., p. 15, Pl. 7, f. 3–4. Lias.; Chile.
M. SOCORRINA, d'Orb., A. Mer., p. 90, Pl. 18, f. 18.
M. orsocrina, d'Orb. (err. typ.), Foss. Col., p. 56. Cret.; Col.
MYOCONCHA ENIGMATICA, Hup. (sp.).
Cardita id., Huppé, Gay's Chile, p. 320, Pl. 5, f. 6. Jur.; Chile.
M. MYTILOIDES, Hup. (sp.).
Cardita id., Hup., Gay, p. 321. Jur.; Chile.
HIPPOPODIUM VALENCIENESII, B. & C. (sp.).
Cardita id., B. & C., M. G. S. Fr., 2 s., v. 4, p. 16, Pl. 6, f. 1-2. Jur.; Chile.
LITHODOMUS SOCIALIS, d'Orb., A. Mer., p. 91. Cret.; Chile.
PTERIA INCONSPICUA, Gabb, new sp. Lias.; Peru.
PERNA AMERICANA, Forbes, Darwin's S. A., p. 266, Pl. 5, f. 4–6. Jur.; Chile.
INOCERAMUS LUNATUS, Forbes, Quart. J. G. Soc., v. 1, p. 179. Cret.; Col.
I. PLICATUS, d'Orb., A. Mer., p. 91, Pl. 18, f. 19. Cret.; Col., Venez.
  Id. Von Buch, Zeits. D. G. Gesell., v. 2, p. 344.
? I. ROEMERI, Karst., G. Verh. N. Gren., p. 112, Pl. 5, fig. 6. Cret.; Col.
  Probably not an *Inoceramus*.

320   DESCRIPTION OF A COLLECTION OF FOSSILS,

TRIGONIA ABRUPTA, V. Buch, Petr., p. 17, f. 21-22.
    Id.   d'Orb., Am. Mer., p. 86, Pl. 19, f. 4-6.   Cret.; Col.
T. BRONNI, Agas. Etud., p. 18, Pl. 5, f. 19.
Lyrodon clavellatum, Bronn, Leth., Pl. 20, f. 3.
Not *T. id.*, Sby., M. Con., Pl. 87.
T. *Bronni*, Gabb, present paper.   Jurassic; Peru.
T. CATENIFERA, Hup., Gay's Chile, p. 328, Pl. 5, f. 8.
T. *catenata* on plate.   Jurassic; Chile.
T. DELAFOSSEI, Bayle & Coq, M. G. Soc. Fr., 2 s., v. 4, p. 37, Pl. 8, f. 27.
    Id.   Hup, Gay, p. 328.   Cret.; Chile.
? T. DOMEYKANA, Phil., Atacama, p. 144, Pl. 1, f. 5-6.   Jur.; Chile.
T. HANETIANA, d'Orb., Am. Mer., p. 127, Pl. 12, f. 14-16.
    Id.   d'Orb., Ast. & Zel., Pl. 5, f. 23-24.
    Id.   Hup., Gay, p. 327.
T. *obtusa*, Hup., *id.*, p. 327, Pl. 5, f. 9.   Cret.; Chile.
T. GIBBOXIANA, Lea, Tr. A. P. S., 2 s, v. 7, p. 255, Pl. 9, f. 7.
T. *Hondaana*, Lea, *loc. cit.*, p. 256, Pl. 9, f. 9.
    Id.   d'Orb., A. Mer., p. 85, Pl. 19, f. 1-3.
T. *Boussingaultii*, d'Orb., Foss. Col., Pl. 4, f. 1-9.
T. *Hondaana*, d'Orb., *loc. cit.*, p. 50.
Not *T. Gibboniana*, Gabb, Pal. Cal., v. 1, p. 190, Pl. 17, f. 178; Pl. 31, f. 262.
Nor *id.*   Gabb, *loc. cit.*, vol. 2, p. 248.   Cret.; Colombia.
    I have compared Mr. Lea's types of the two names and find they are undoubtedly the same species, and on comparing them with the Californian form, they prove to be quite distinct.
T. HUMBOLDTII, V. B., Petr., p. 9, f. 28-30.   Cret.; Col
T. LAJOYI, Desh., Leym., M. G. S. Fr., v. 5, Pl. 8, f. 4.
    Id.   d'Orb., A. Mer., p. 87, Pl. 19, f. 10-11.
T. *longa*, Agas., Etud. Trigon., No. 47, Pl. 8, f. 1.   Cret.; Col.
T. LORENTI, Dana, Wilkes' Exped., p. 721, Pl. 15, f. 2.
    Id.   G., present paper.   Jur.; Peru.
T. SUBCRENULATA, d'Orb., A. Mer., p. 87, Pl. 19, f. 7-9.   Cret.; Col
T. TOCAIMAANA, Lea, Trans. A. P. S., 2 s., v. 7, p. 256, Pl. 9, f. 8.
T. *alæformis*, Von Buch, Petr., p. 8, Pl. 1, f. 10.
    Id.   d'Orb., A. Mer., p. 88, Pl. 20, f. 1.
Not *id.* Sby., M. Con., Pl. 3, f. 27.
    I had coincided with the two above-quoted authors in believing this shell to be the same as the English form until, since I have been at work at the present paper, Mr. Lea kindly loaned me his specimens, and I find, on a critical comparison with authentic English specimens, that there are small, though constant specific differences.
MYOPHORIA SPINALIS, Gabb, new sp.   Lias.; Peru.
POSIDONOMYA BECHERI, var. *Liasana*, Bronn, Leth., p. 342, Pl. 18, f. 23.
    Id.   Phil., Atacama, p. 144, Pl. 1, fig. 7.   Lias.; Chile.
ARCA ARAUCANA, d'Orb., A. Mer., p. 129, Pl. 13, f. 1-2.   Cret.; Chile.

MADE BY DR. ANTONIO RAIMONDI IN PERU. 321

A. Orestis, Rath., P. Bost. N. H. Soc., v. 17, p. 247.	Cret.; Brazil.
? A. ovalis, Gabb, new sp.	Cret.; Peru.
A. perodliqua, V. Buch, Petr., p. 16, figs. 13-14.	Cret.; Col.
A. rostellata, V. Buch, id., p. 14, f. 16.	Cret.; Col.
A. Santiaguensis, Hup., Gay, p. 300, Pl. 5, f. 10.	
A. Huidobrii on plate.	Jur.; Chile.
A. Tocaymensis, d'Orb., Prod., v. 2, p. 107.	
Cucullæa id., d'Orb., A. Mer., p. 90, Pl. 21, f. 1-3.	Cret.; Col.
Barbatia Raimondii, Gabb, new sp.	Lias.; Peru.
Cucullæa (Trigonarca?) brevis, d'Orb., A. Mer., p. 89, Pl. 20, f. 2-4.	
Arca id., d'Orb., Prod., v. 2, p. 106.	
C. id., Gabb, present paper.	Cret.; Peru, Col.
C. Gabrielis, Leym., M. G. S. Fr., v. 5, Pl. 7, f. 5.	
Arca id., d'Orb., Prod., v. 2, p. 80.	
C. dilatata, d'Orb., A. Mer., p. 89, Pl. 20, f. 5-7.	Cret.; Col.
Cucullæa Hartii, Rath.	
Arca (Cuc.) id., Rath., P. B. N. H. Soc., v. 17, p. 249.	Cret.; Brazil.
C. (Trigonarca?) Ordignyana, Gabb, new sp.	Cret.; Peru.
C. (Trigonarca) Peruana, Gabb, new sp.	Cret.; Peru.
C. subcentralis, Rath., P. B. N. H. Soc., v. 17, 249.	Cret.; Brazil.
Nucula? Albertina, d'Orb., Ast. & Zel., Pl. 5, f. 25-26.	Cret.; Chile.

This shell, evidently described without seeing the hinge, looks to me more like one of the *Veneridæ*, than a *Nucula*.

N. Mariæ, Rath., Proc. Bost. N. H. Soc., v. 17, p. 214.	Cret.; Brazil.
N. Peruana, Gabb, new sp.	Lias.; Peru.
? N. incerta, d'Orb., A. Mer., p. 85.	
Id. d'Orb., Foss. Col., p. 50.	Cret.; Col.
Nuculana Swiftiana, Rath. (sp.).	
Leda id., Rath., Proc. Bost. N. H. Soc., v. 17, p. 245.	Cret.; Brazil.
N. Braziliana, Rath. (sp.).	
Leda id., Rath., loc. cit., p. 246.	Cret.; Brazil.
Pecten abnormis, Hup., Gay's Chile, p. 292, Pl. 5, f. 3.	Jur.; Chile.
P. argentarius, Gabb, new sp.	? Jur.; Peru.
"? P. (Terebratula?)" Deserti, Phil.. Atacama, p. 145, Pl. 1, f. 9.	Jur.; Chile.
Pecten granulatus, d'Orb., Ast. & Zel., Pl. 5, f. 29-30.	Cret.; Chile.
P. Raimondii, Gabb, new sp.	Jur.? Chile.
P. unguiferus, Hup., Gay's Chile, p. 292, Pl. 5, f. 1.	Jur.; Chile.
Pleuronoctes Chilensis, d'Orb. (sp.).	
Pecten id., d'Orb., Ast. & Zel., Pl. 5, f. 31, 32.	Cret.; Chile.
Neithea alata, V. Buch (sp.).	
Pecten alatus, V. B., Petr., p. 3, Pl. 1, f. 1-4.	
P. Dufrenoyi, d'Orb., A. Mer., p. 106, Pl. 22, f. 5-9.	
Id. Hup., Gay's Chile, p. 291.	

P. alatus, Bayle & Coq., M. G. Soc. Fr., 2 s., v. 4, p. 14, Pl. 5, f. 1-2.
    Id.   Con., U. S. Astron. Exped., p. 283, Pl. 41, f. 2.
Janira id., Rem., Pal. Chile, p. 18.
Neithea id., G., Syn. Cret., p. 147.   From Colombia to Chile.

The age of this fossil has been the cause of great difference of opinion. Dr. Raimondi calls his specimens Jurassic, with some doubt. D'Orbigny places it in the Neocomien, while others carry it as low as the Lias. If it is older than Cretaceous it is the only species of the genus that is found outside of the limits of that formation. My own opinion is that it is somewhat high in the Cretaceous; not only from the fact that the genus is eminently characteristic of the chalk and green sand; but because all of the other fossils, sent with it by Humboldt and Degenhardt from Colombia to Von Buch, as well as the additional species collected by Dr. Gibbon and described by Dr. Lea, and, again the third collection, described from the same region by Forbes, all seem to belong to the age of the white chalk, or not far below it.

NEITHEA QUINQUECOSTATA, Sby. (sp.).
Pecten id., Sby., M. Con., Pl. 56.
Janira id., d'Orb., P. Fr., T. Cret., p. 633, Pl. 414, f. 1-5.
Neithea id., Gabb, Syn. Cret., 1861, p. 148.
    Id.   Gabb, present paper.   Cret.; Peru.
LIMA RARICOSTA, Bayle & Coq., M. G. S. Fr., 2 s., v. 4, p. 26, Pl. 6, f. 3-4.   Jur.; Chile.
L. (*Plagiostoma*) TRUNCATIFRONS, B. & C., *loc. cit.*, p. 25, Pl. 6, f. 5.
    Id.   Hup., Gay's Chile, p. 296.   Jur.; Chile.
L. (*P.*) DUBIA, Hup., Gay, p. 297, Pl. 9, f. 5.   Jur.; Chile.
PLICATULA RAPA, B. & C., M. G. S. Fr., 2 s. v. 4, p. 16, Pl. 5, f. 8-10.
    Id.   Hup., Gay, p. 293.   Jur.; Chile.
P. TORTA, Gabb, new sp.   Cret.; Peru.
ANOMIA PARVA, Gabb, Proc. Phil. Acad., 1860, p. 198, Pl. 3, f. 15.   Cret.; Chile.
A. PERUANA, Gabb, new sp.   Cret.; Peru.
PLACUNANOMIA LIMA, Gabb, new sp.   Cret.; Peru.
OSTREA ADRUPTA, d'Orb., A. Mer., p. 93, Pl. 21, f. 4-6.   Cret.; Col.
O. (*Exogyra?*) ATACAMENSIS, Phil., Atacama, p. 145, Pl. 1, f. 11-12.   Jur.? Chile.
O. CALLACTA, Con., Proc. Acad. N. S., Phil., 1875, p. 139, Pl. 22, f. 1.
O. sp. indet., Dana, Wilkes' Exped., Pl. 15, f. 7.
O. *callacta*, Gabb, present paper.   Jur.; Peru.
O. EXCARPIFERA, Hup., Gay's Chile, p. 286.   Jur.; Chile.
O. GREGARIA, Sby., M. Con., Pl. 111, f. 1-3.
    Id.   Bayle & Coq., M. G. S. Fr. 2 s., v. 4, p. 24.
    Id.   Hup., Gay's Chile, p. 285.
    Id.   Con., U. S. Astron. Exped., p. 283, Pl. 41, f. 1.   Jur.; Chile.
O. INOCERAMOIDES, d'Orb., A. Mer., p. 94.   Cret.; Col.
O. LARVIFORMIS, Gabb, new sp.   Cret.; Peru.
O. MARSHII, Sby., M. Con., Pl. 48.
    Id.   B. & Coq., M. G. Soc. Fr, p. 21, Pl. 5, f. 3.
    Id.   Hup., Gay's Chile, p. 284.   Jur.; Chile.

O. PULLIGERA, Goldf., Petr., v. 2, Pl. 72.
O. *solitaria*, Sby., M. Con., Pl. 468, f. 1.
O. *pulligera*, B. & C., M. G. Soc. Fr., 2 s., v. 4, p. 21, Pl. 5, f. 4–5.
   Id.   Hup., Gay, p. 283.   Jur.; Chile.
O. RIVOTI, B. & Coq., M. G. S. Fr., 2 s., v. 4, p. 24, Pl. 1, f. 7–8.
O. *cymbium*, B. & C. (*not Lam.*), *loc. cit.*, Pl. 5, f. 6–7.
O. *irregularis*, Con. (*not Munst.*), Astr. Exped., p. 283, Pl. 42, f. 9.
O. *Rivoti*, Hup., Gay's Chile, p. 284.   Jur.; Chile, Peru.
O. SANDALINA, Goldf., Petr., Pl. 79, f. 9.
   Id.   B. & Coq., Huppé.   Jur.; Chile.
O. SANTIAGUENSIS, Hup., Gay's Chile, p. 288, Pl. 3, f. 3.
O. *transitoria*, on plate.   Jur.; Chile.
GRYPHÆA ARCUATA, Lam., Syst. Nat., Pl. 6, p. 198.
   Id.   Goldf., Petr. p. 28, Pl. 84, f. 1.
G. *incurva*, Sby., M. Con., Pl. 112.
G. *cymbium*, Schlot., not Lam.
   Id.   Phil., Atacama, p. 144.   Jurassic; Chile.
This probably includes also *G. Darwini*, Forbes, *q. v.*
G. CYMBIUM, Lam., Syst., vol. 6, p. 198.
   Id.   B. & C., M. G. S. Fr., 2 s., v. 4, p. 13, Pl. 4, f. 1.
Ostrea hemispherica, d'Orb., A. Mer., p. 106, Pl. 22, f. 3–4.
   cymbium, Hup., Gay's Chile, p. 287.   Lias.; Peru, Chile.
G. DARWINI, Fbs., Darwin's S. A., p. 266, Pl. 5, f. 7.   "Secondary;" Jur.? Chile.
Probably = *G. arcuata*.
G. DILATATA, Sby., M. Con., Pl. 149, f. 2–3.
Ostrea id., Phil., Atacama, p. 144.   Jur.; Chile.
"OSTREA (*Gryphæa?*) STRIATA," Phil., Atacama, p. 144, Pl. 1, f. 10.   Jur.; Chile.
G. VESICULOIDES, Gabb, new sp.   Cret.; Peru.
EXOGYRA LATERALIS, Nils. (sp.), Dub., Bull. G. Soc. Fr., vol. 8, p. 385.
Ostrea id., Nils., Pal. Suec., p. 29, Pl. 7, f. 10.
G. *vomer*, Mort., Syn. Cret., p. 54, Pl. 9, f. 5.
E. *lateralis*, Rath., Proc. Bost. N. H. Soc., v. 17, p. 243.   Cret.; Brazil.
E. OBLONGA, Huppé (sp.).
Ostrea id., Hup., Gay's Chile, p. 284, Pl. 4, f. 2.   Jur.; Chile.
E. PARASITICA, Gabb, Pal. Cal., v. 1, p. 205, Pl. 26, fig. 192; Pl. 31, f. 273.
   Id.   Gabb, present paper.   Cret.; Peru
E. PLICATA, Lam. sp., Goldf., Petr., p. 37, Pl. 87, f. 5.
E. *Boussingaultii*, d'Orb., A. Mer., p. 91, Pl. 18, f. 20; Pl. 20, f. 8–9.   Cret.; Peru, Col.
For further synonymy of this species, see Palæontology of California, v. 2, p. 275.
E. POLYGONA, Von Buch, Petr., p. 5, figs. 8–9.
Ostrea Couloni, B. & Coq., M. G. S. Fr., 2 s., v. 4, p. 37, Pl. 7, f. 1–2.
Exogyra id., d'Orb. (not Defr.), A. Mer., p. 93.
E. *polygona*, Gabb, present paper.   Cret.; Colombia to Chile.

E. squamata, d'Orb., A. Mer., p. 92, Pl. 19, f. 12-15.
Id.   d'Orb., Foss. Col., p. 58, Pl. 4, f. 12-15.
Ostrea subsquamata, d'Orb., Prod., v. 2, p. 108. Cret.; Colombia.
Hippurites Chilensis, d'Orb., A. Mer., p. 107, Pl. 22, f. 16. Cret.; Chile.

BRACHIOPODA.

Terebratula bicanaliculata, Schlot., in Ziet., Pl. 40. f. 5.
    Id.   B. & Coq., M. G. S. Fr., 2 s., v. 4, p. 31, Pl. 8, f. 17-19.
    Id.   Hup., Gay, p. 406. Jur.; Chile.
T. Domeykanus, B. & Coq., M. G. S. Fr., 2 s., v. 4, p. 30, Pl. 8, f. 1-3.
    Id.   Hup., Gay, p 403. Jur.; Chile.
T. emarginata, Sby., M. Con., Pl. 435, f. 5.
    Id.   B. & Coq., M. G. S. Fr., 2 s., v. 4, p. 32, Pl. 9, f. 7-9.
    Id.   Hup., Gay, p. 406. Jur.; Chile.
T. ficoides, B. & Coq., M. G. S. F., 2 s., v. 4, p. 30, Pl. 8, f. 20-22.
    Id.   Hup., Gay, p. 405. Jur.; Chile.
T. Haueri, Karst., Geog. Verh. N. Gren., p. 113, Pl. 6, f. 1. Cret.; Col.
T. meridionalis, Con., U. S. Astron. Exped., p. 282, Pl. 42, f. 10. Cret.? Chile.
T. ornithocephala, Sby., M. Con., Pl. 101.
T. Ignaciana, Fbs., Darwin's S. A., p. 63, Pl. 22, f. 14-15.
T. ornithocephala, B. & Coq., M. G. S. Fr., 2 s., v. 4, p. 18, Pl. 3, f. 12-14.
    Id.   Hup., Gay, p. 404. Jur.; Chile.
T. perovalis, Sby., M. Con., Pl. 436, f. 2-3.
T. Inca, Fbs., Darwin's S. A., p. 268, Pl. 5, f. 19-20.
T. perovalis, B. & Coq., M. G. S. Fr., 2 s., v. 4, p. 28, Pl. 8, f. 15-16.
    Id.   Hup., Gay, p. 403. Jur.; Peru, Chile.
T. Raimondiana, Gabb, new sp. Jur.? Peru.
T. subexcavata, Con., U. S. Astron. Exped., p. 282, Pl. 41, f. 4. Cret.; Chile.
Rhynchonella Antonii, Gabb, new sp. Jur.; Peru.
R. concinna, Sby. (sp.), d'Orb., Prod., vol. 1, p. 315.
Terebratula id., Sby., M. Con., p. 189, Pl. 83, f. 6-7.
T. ænigma, d'Orb., A. Mer., p. 62, Pl. 22, f. 10-13.
T. concinna, B. & Coq., M. G. S. Fr., 2 s., v. 4, p. 23, Pl. 9, f. 4-6.
    Id.   Hup., Gay's Chile, p. 405. Jur.; Chile.
R. lacunosa, Schlot. (sp.), d'Orb., Prod., p. 375.
Terebratula id., Schlot., Petr., Pl. 1, f. 2.
R. id., B. & Coq., M. G. S. Fr., 2 s., v. 4, p. 29, Pl. 9, f. 10-11. Jur.; Chile.
R. subtetrahedra, Con. (sp.).
Terebratula id., Con., U. S. Astron. Exped., p. 282, Pl. 42, f. 8. Jur.; Chile.
R. tetrahedra, Sby. (sp.), d'Orb., Prod.
T. id., Sby., M. Con., Pl. 83, f. 4.
T. id., B. & Coq., M. G. S. Fr., 2 s., v. 4, p. 17, Pl. 7, f. 9-10.
T. id., Hup., Gay's Chile, p. 404. Jur.; Chile.

SPIRIFER TUMIDUS, Buch., M. G. S. Fr., vol. 4, Pl. 10, f. 20.
S. *Chilensis*, Fbs., Darwin's S. A., p. 267, Pl. 5, f. 15, 16.
S. *linguiferoides*, Fbs., *loc. cit.*, figs. 17–18.
S. *tumidus*, Hup., Gay's Chile, p. 407.   Lias.; Chile.

RADIATA.
ECHINODERMATA.

ECHINUS ANDINUS, Phil., Atacama, p. 146, Pl. 2, f. 11–13.	Jur.; Chile.
E. DIGRANULARIS, Lam., B. & Coq., M. G. S. Fr., v. 4, p. 32, Pl. 8, f. 35–36.	Jur.; Chile.
E. BOLIVARII, d'Orb., A. Mer., p. 95, Pl. 21, f. 11–13.	
Id. d'Orb., Foss. Col., p. 61, Pl. 6, f. 11–13.	
Id. Gabb, present paper.	Cret.; Col. & Peru.
E. DIADEMOIDES, B. & Coq., *loc. cit.*, p. 33, Pl. 8, f. 33, 34.	Jur.; Chile.
CIDARITES OVATA, Phil., Atacama, p. 146, Pl. 1, f. 13–14.	Jur.; Chile.
PYGURUS COLUMBIANUS, d'Orb., Prod., v. 2, p. 109.	
Laganum id., d'Orb., A. Mer., p. 95, Pl. 21, f. 10.	
Id. d'Orb., Foss. Col., p. 60, Pl. 6, f. 10.	Cret.; Col.
BOTRIOPYGUS ELEVATUS, Gabb, new sp.	Cret.; Peru.
B. COMPRESSUS, Gabb, new sp.	Cret.; Peru.
DISCOIDEA EXCENTRICA, d'Orb., A. Mer., p. 94, Pl. 21, f. 7–9.	
Id. d'Orb., Foss. Col., p. 60, Pl. 6, f. 7–9.	Cret.; Col.
D. NUMISMALIS, Gabb, new sp.	Cret.; Peru.
ENNALASTER PERUANA; Gabb, new sp.	Cret.; Peru.
PERIASTER AUSTRALIS, Gabb, new sp.	Cret.; Peru.
MICRASTER CHILENSIS, Phil., Atacama, p. 147, Pl. 2, f. 8–10.	Jur.; Chile.
? SPATANGUS COLOMBIANUS, Lea, Tr. Amer. P. Soc., 2 s., v. 7, p. 257, Pl. 9, f. 11.	Cret.; Col.

FORAMINIFERA.

CYCLOPÆA RUMICHACÆ, Karst., Geog. Verh. N. Gren., p. 113, Pl. 6, f. 2.	Cret.; Col.
ORTHOCERINA EWALDI, Karst., *loc. cit.*, p. 114, Pl. 6, f. 3.	Cret.; Col.
PLANULINA ZAPATOCENSIS, Karst., *loc. cit.*, p. 114, Pl. 6, f. 4.	Cret.; Col.
RODULINA SOGAMOZÆ, Karst., *loc. cit.*, p. 114, Pl. 6, f. 5.	Cret.; Col.
ORBITULITES LENTICULARIS, Karst., *loc. cit.*, p. 114, Pl. 6, f. 6.	Cret.; Col.

CAINOZOIC.
CRUSTACEA.
CIRRIPEDIA.

BALANUS LÆVIS, Brug., Enc. Meth., Pl. 164, f. 1.
B. *discors*, Ranzoni, Mem. di Storia Nat. 1820, Pl. 3, f. 9–13.
B. *Coquimbensis*, Sby., Darwin's S. A., p. 264, Pl. 2, f. 7.
B. *lævis*, Darwin, Monog. Cirr., p. 227, Pl. 4, f. 2.   Tert.; Chile & Patagonia.

B. PSITTACUS, Molina (sp.), King & Brod., Zool. Jour., v. 5, p. 332.
Lepas id., Molina, N. H. Chile, v. 1, p. 223.
B. picos, Lesson, Voy. Coquille.
B. tintinabulum (var. C.), Ranzani, Mem. Storia N., 1820, Pl. 3, f. 1-3.
B. cylindraceus, Lam., Chenu Conch. Ill., Pl. 4, f. 17; Pl. 5, f. 7.
Not id., Lam., A. S. Vert.
B. psittaceus, Darwin, Monog. Cirr., p. 206, Pl. 2, f. 3.   Tert.; Chile, Peru.
B. VARIANS, Sby., Darwin's S. A., p. 264, Pl. 2, f. 4-6.
   Id. Darwin, Monog. Cirr., p. 298, Pl. 8, f. 9.   Tert.; Chile, Peru.

MOLLUSCA.

GASTEROPODA

TROPHON PATAGONICUS, Sby. (sp.).
Fusus id., Sby., Darwin's S. A., p. 259, Pl. 4, f. 60.   Tert.; Patagonia.
? FUSUS PYRULIFORMIS, Sby., Darwin's S. A., p. 258, Pl. 4, f. 56.   Tert.; Chile.
Not a true Fusus.
F. PAYTENSIS, Gabb, J. Conch., 1869, p. 25.
   Id. Gabb, present paper.   Tert.; Peru.
FUSUS (*Volema*) SUBREGULARIS, d'Orb. (sp.).
F.? regularis, Sby., Darwin's S. A., p. 258, Pl. 4, f. 55.
Not id., Desh.
F. subregularis, d'Orb., Prod., v. 3, p. 69, No. 1262.
F. striato-nodosus, Hup., Gay's Chile, p. 174, Pl. 2, f. 5.   Tert.; Chile.
PRISCOFUSUS SUBREPLEXUS, Sby. (sp.).
Fusus id., Sby., Darwin's S. A., p. 259, Pl. 4, f. 57.   Tert.; Chile.
NEPTUNEA CLATHRATA, Huppé (sp.).
Fusus id., Hup., Gay, p. 174, Pl. 2, f. 4.   Tert.; Chile.
N. CLERYANA, d'Orb. (sp.).
Fusus id., d'Orb., A. Mer., p. 117, Pl. 12, f. 6-9.   Tert.; Chile.
   Id. Hup., Gay's Chile, p. 172.
   Rémond has referred this species to the Cretaceous, but the other authors agree in calling it Tertiary.
N. NOACHINA, Sby. (sp.).
Fusus Noachinus, Sby., Darwin's S. A., p. 259, Pl. 4, f. 58, 59.   Tert.; Pat.
N. ORDIGNYI, Hup. (sp.).
Fusus id., Huppé, Gay's Chile, p. 175, Pl. 3, f. 5.
F. sulcatus on plate.   Tert.; Chile.
N. PETITIANA, d'Orb. (sp.).
Fusus id., d'Orb., A. Mer., p. 118, Pl. 12, f. 10.   Tert.; Chile.
CASSIDULA ECHINULATA, Hup. (sp.).
Fusus id., Hup. (sp.), Gay's Chile, p. 173, Pl. 2, f. 3.   Tert.; Chile.
CLAVELLA SOLIDA, Nelson, Tr. Conn. Acad., v. 2, p. 199.   Tert.; Peru.

PLEUROTOMA TURBINELLOIDES, Sby., Darwin's S. A., p. 258, Pl. 4, f. 53.   Tert.; Chile.
Evidently not a *Pleurotoma* (or better *Turris*) in the restricted sense, but we have not sufficient information to refer the species to its proper genus.
SURCULA DISCORS, Sby. (sp.).
Pleurotoma id., Sby., Darwin, p. 258, Pl. 4, f. 54.   Tert.; Chile.
MANGELIA LANCEOLATA, Hup. (sp.).
Pleurotoma id., Hup., Gay's Chile, p. 117, Pl. 3, f. 7.   Tert.; Chile.
TRITONIUM ARMATUM, Hup. (sp.).
Triton id., Hup., Gay's Chile, p. 182, Pl. 3, f. 1.   Tert.; Chile.
T. PERNODOSUM, Gabb, present paper.   Tert.; Peru.
T. VERRUCULOSUM, Sby. (sp.).
Triton id., Sby., Darwin's S. A., p. 260, Pl. 4, f. 63.   Tert.; Chile.
T. (*Ranularia*) LEUCOSTOMOIDES, Sby.
Triton id., Sby., Darwin, p. 260, Pl. 4, f. 64.   Tert.; Chile.
T. (*Argobuccinum*) SCABER, Brod.
Triton id., Brod., Zool. Jour., 1822, p. 348.
    Id.   d'Orb., A. Mer., p. 158.   Quart.; Bolivia.
T. (*A.*) ZORRITENSE, Nelson.
Argobuccinum id., N., Tr. Conn. Acad., v. 2, p. 198, Pl. 7, f. 1-2.   Tert.; Peru.
BUCCINANOPS COCHLIDIUM, Chenu (sp.), d'Orb., Moll. A. Mer., p. 354, Pl. 61, f. 25.
    Id.   d'Orb., Foss. A. Mer., p. 157.
Buccinum id., Chenu, Conch. Cab., v. 11, p. 275, Pl. 209, f. 2053.   Quart.; Pat.
B. GLOBULOSUM, Kien. (sp.), d'Orb., Moll. A. Mer., p. 355, Pl. 61, f. 24.
    Id.   d'Orb., Foss. A. Mer., p. 157.
Buccinum id., Kiener, Buccinum, p. 12, Pl. 10, f. 33.   Quart.; Peru, Arg. Rep.
PURPURA CASSIDIFORMIS, Blainv., M. sur Purp., p. 42.
    Id.   Hup., Gay's Chile, p. 188.   Quart.; Chile.
P. CHOCOLATA, Ducl., A. Sc. Nat., v. 26, Pl. 2, f. 7.
    Id.   d'Orb., Foss. A. Mer., p. 157.   Quart.; Bol.
    Id.   Gabb, Am. Jour. Conch., 1869, p. 26.   Tert.; Peru.
CONCHOLEPAS PERUVIANUS, Lam., A. S. V.
Purpura concholepas, d'Orb., Moll. A. Mer., p. 360, Pl. 61, f. 5-7.
    Id.   d'Orb., Foss. A. Mer., p. 158.
    Id.   Hup., Gay, p. 201.   Quart.; Chile; Bol.
C. KIENERI, Hup., Gay, p. 203, Pl. 3, f. 4.   Quart.; Chile.
GASTRIDIUM CEPA, Sby., Darwin, p. 261, Pl. 4, f. 68-69.   Tert.; Chile.
MONOCERAS AMBIGUUS, Sby., Darwin, p. 261, Pl. 4, f. 66-67.   Tert.; Chile.
M. BLAINVILLEI, d'Orb., A. Mer., p. 116, Pl. 6, f. 18-19.
    Id.   Hup., Gay, p. 197.   Tert.; Peru? Chile.
M. GIGANTEUM, Less., Voy. Coq., p. 405, Pl. 11, f. 4.
    Id.   Hup., Gay, p. 198.   Tert.; Chile.
M. LABIALE, Hup., Gay, p. 199, Pl. 3, f. 3.   Tert.; Chile.
M OPIMUM, Hup., Gay, p. 200, Pl. 2, f. 6.   Tert.; Chile.

M. UNICORNE, Brug. (sp.).
Buccinum id., Brug., Enc. Meth., t. 396, f. 2.
M. crassilabrum, Lam., A. S. V.
M. unicorne, Gray, etc., Hup., Gay's Chile, p. 194. — Quart.; Chile.
CUMA ALTERNATA, Nelson, Tr. Conn. Acad., v. 2, p. 198, Pl. 7, f. 3-4. — Tert.; Peru.
?OLIVA SERENA, d'Orb., A. Mer., p. 116, Pl. 14, f. 9 — Tert.; Chile.
PORPHYRIA PERUVIANA, Lam. (sp.), Auct.
Oliva id., Lam., A. S. V.
   Id.   Hup., Gay's Chile. p. 216. — Quart.; Chile.
OLIVELLA DIMIDIATA, Sby. (sp.).
Oliva id., Sby., Darwin. p. 263, Pl. 4, f. 76-77. — Tert.; Chile.
O. SIMPLEX, Hup. (sp.).
Oliva id., Hup., Gay's Chile, p. 217. Pl. 3, f. 9. — Tert.; Chile.
OLIVANCILLARIA AURICULARIS, Lam. (sp.), d'Orb., Moll. A. M., p. 421, Pl. 59, f. 20-22.
   Id.   d'Orb., Foss. A. Mer., p. 156.
Oliva id., Lam., A. S. V., v. 7, p. 434. — Quart.; Pat.
O. BRAZILIENSIS, Chenu (sp.), d'Orb., M. A. Mer., p. 420.
   Id.   d'Orb., Foss. A. Mer., p. 155.
Oliva id., Chenu, Conch. Cab., p. 130, Pl. 147, f. 1367-1370. — Quart.; Pat.
O. TUMORIFERA, Hup. (sp.).
Oliva id., Hup., Gay, p. 217, Pl. 3, f. 8. — Tert.; Chile.
FASCIOLARIA TRIPLICATA, Sby. (sp.).
Voluta id., Sby., Darwin, p. 262, Pl. 4, f. 74.
V. subtriplicata, d'Orb., Prod., v. 3, p. 53. — Tert.; Chile.
CYMBIOLA ALTA, Sby. (sp.).
Voluta id., Sby., Darwin, p. 262, Pl. 4, f. 75. — Tert.; Chile, Pat.
C. BRAZILIANA, Solander (sp.), H. & R. Ad. Genera.
Voluta id., Sol., Cat. Portl. Mus., No. 3958.
   Id.   Lam., A. S. V., v. 7, p. 355.
   Id.   d'Orb., A. Mer., p. 156. — Quart.; Pat., Arg. Rep.
VOLUTELLA ANGULATA, Sw. (sp.)?
Voluta id., Swains., Donovan, Pl. 1.
Volutella id., d'Orb., A. Mer., Moll., p. 423, Pl. 60, f. 1-3.
   Id.   d'Orb., Foss. A. Mer., p. 156. — Quart.; Pat.
SCAPHELLA TUBERCULATA, Wood (sp.).
Voluta id., Wood, Ind. Supp., No. 22.
   Id.   d'Orb., A. Mer., p. 157. — Quart.; Pat.
VOLUTIDERMA PLICIFERA, Gabb, present paper.
Volutilithes id., Gabb, Am. Jour. Conch., 1869, p. 28. — Tert.; Peru.
MARGINELLA INCRASSATA, Nelson, Tr. Conn. Acad., v. 2, p. 197, Pl. 6, f. 5-6. — Tert.; Peru.
STROMBINA LANCEOLATA, Sby. (sp.), Chr., Br. Assn. Rep., 1856.
   Id.   Nelson, Tr. Conn. Acad, v. 2, p. 197.
Columbella id., Sby., Pr. Zool. Soc., 1832, p. 116. — Tert.; Peru.

MADE BY DR. ANTONIO RAIMONDI IN PERU. 329

GALEODEA MONILIFERA, Sby. (sp.).
Cassis id., Sby., Darwin, p. 260, Pl. 4, f. 65.   Tert.; Chile.
G. TUBERCULIFERA, Hup. (sp.).
Cassidaria id., Hup., Gay's Chile, p. 209, Pl. 3, f. 2.   Tert.; Chile.
FICUS DISTANS, Sby. (sp.).
Pyrula id., Sby., Darwin, p. 259, Pl. 4, f. 61.   Tert.; Chile.
EUSPIRA ORTONI, Gabb. present paper.
Ampullina id., G., Am. Jour. Conch., 1869, p. 27.   Tert.; Peru.
LUNATIA ISABELLINA, d'Orb. (sp.).
Natica id., d'Orb., A. Mer., Moll., p. 402, Pl. 76, f. 12–13.
   Id.   d'Orb., Foss. A. M., p. 154.   Quart.; Arg. Rep.
L. LIMBATA, d'Orb. (sp.).
Natica id., d'Orb., Moll. A. Mer., p. 402, Pl. 57, f. 7–9.
   Id.   d'Orb., Foss. A. M., p. 154.   Quart.; Pat.
NEVERITA PACHYSTOMA, Hup. (sp.).
Natica id., Hup., Gay's Chile, p. 223, Pl. 1, f. 6.   Tert.; Chile.
UBER SUBANGULATA, Nelson (sp.).
Polinices id., Nelson, Tr. Conn. Acad., v. 2, p. 195, Pl. 6, f. 4, 12–13.   Tert.; Peru.
SIGARETUS SUBGLOBOSUS, Sby., Darwin, p. 254, Pl. 3, f. 36–37.   Tert.; Chile.
S. ELEGANS, Hup., Gay, p. 226, Pl. 1, f. 5.   Tert.; Chile.
SCALARIA (*Clathrus*) ELEGANS, d'Orb.
S. elegans, d'Orb., Moll. A. Mer., p. 389, Pl. 54, f. 1–3.
   Id.   d'Orb., Foss. *id* , p. 154.   Quart.; Pat., Arg. Rep.
S. (*C.*) RUGULOSA, Sby.
S. rugulosa, Sby., Darwin, p. 255, Pl. 3, f. 42–43.   Tert.; Pat.
TURBONILLA MINUSCULA, Gabb, J. Conch., 1868, p. 197, Pl. 16, f. 1.
   Id.   Gabb, present paper.   Tert.; Peru.
TEREBRA COSTELLATA, Sby., Darwin, p. 262, Pl. 4, f. 70–71.   Tert.; Chile.
T. (*Myurella*) TUBEROSA, Nelson.
M. tuberosa, Nelson, Tr. Conn. Acad., v. 2, p. 193.   Tert.; Peru.
T. UNDULIFERA, Sby., Darwin, p. 262, Pl. 4, f. 72–73.   Tert.; Chile.
ARCHITECTONICA COLLARIS, Sby. (sp.).
Trochus id., Sby., Darwin, p. 256, Pl. 3, f. 44–45.   Tert.; Chile.
A. LÆVIS, Sby., *id.*, p. 256. Pl. 3, f. 46–47.   Tert.; Chile.

These two species, belonging to the same genus or subgenus, differ from the typical form of *Architectonica* in the absence of the characteristic sculpture; they probably indicate a new subgenus.

A. SEXLINEARIS, Nelson (sp.).
Solarium id., N., Tr. Conn. Acad., v. 2, p. 194, Pl. 6, f. 11.   Tert.; Peru.
RIMELLA GAUDICHAUDI, d'Orb. (sp.).
Rostellaria id., d'Orb., A. Mer., p. 116, Pl. 14, f. 6–8.   Tert.; Peru.
CANCELLARIA BRADLEYI, Nelson, Tr. Conn. Acad., v. 2, p. 192, Pl. 6, fig. 8–9.   Tert.; Peru.
C. LARKINII, N., *loc. cit.*, p. 192, Pl. 6, f. 7.   Tert.; Peru.

330 DESCRIPTION OF A COLLECTION OF FOSSILS,

C. (*Aphera*) Peruana, N., *loc. cit.*, p. 190, Pl. 6, f. 3.	Tert.; Peru.
C. spatiosa, N., *loc. cit.*, p. 191.	Tert.; Peru.
C. triangularis, Nelson, *loc. cit.*, p. 191, Pl. 6, f. 10.	Tert.; Peru.
Trichotropis (*Iphinoe*) ornata, Shy. (sp.).	
Struthiolaria ornata, Shy., Darwin, p. 260, Pl. 4, fig. 62.	Tert.; Pat.
Cerithium læviusculum, Gabb, J. Conch., 1869, p. 27.	
Id. Gabb, present paper.	Tert.; Peru.
Planorbis Perasana, Con., Proc. Acad. Nat. Sci., Phila., v. 26, p. 30.	Tert.; Brazil.
Paludinestra Australis, d'Orb., Moll. A. Mer., p. 384, Pl. 48, f. 4–6.	
Id. d'Orb., Foss. A. Mer., p. 153.	Quart.; Patagonia.
Liosoma curta, Con., P. Acad. Nat. Sci., Phila., v. 26, p. 31, Pl. 1, f. 8.	Tert.; Brazil.
Isæa lintea, Con., Jour. Conch., v. 6, p. 193, Pl, 10, f. 6.	Tert.; Brazil.
I. Ortoni, Gabb (sp.), Con., J. Conch., v. 6, p. 193, Pl. 10, f. 10–13; Pl. 11, f. 8.	
Mesalia id., Gabb, J. Conch. v. 4, p. 198, Pl. 16, f. 3.	Tert.; Brazil.
Hemisinus Steerei, Con., P. Acad. Nat. Sci., Phila., v. 26, p. 32, Pl. 1, f. 14.	Tert.; Brazil.
H. sulcatus, Con., J. Conch., v. 6, p. 194, Pl. 10, f. 2.	Tert.; Brazil.
Dyris gracilis, Con., J. Conch., p. 195, Pl. 10, f. 8; Pl. 11, f. 7.	Tert.; Brazil.
Enora (*Neseis*) bella, Con., J. C., v. 6, p. 184, Pl. 10, f. 17.	
Fossar bella, Woodw., A. M. N. H, 1871, p. 102, Pl 5, f. 3.	Tert.; Brazil.
E. crassilabra, Con., J. C., v. 6, p. 194, Pl. 10, f. 14.	
Id. Con., P. Acad. Nat. Sci., Phila., v. 26, p. 32, Pl. 1, f. 9.	Tert.; Brazil
Littorina laqueata, Gabb, J. Conch., 1869, p. 28.	
Id. Gabb, present paper.	Tert.; Peru.
Turritella affinis, Hup., Gay, p. 155, Pl. 2, f. 7.	Quart.; Pat.
T. ambulacrum, Shy., Darwin, p. 257, Pl. 3, f. 49.	
T. suturalis, Shy., *loc. cit.*, p. 257, Pl. 3, f. 50.	Tert.; Pat.
T. bifasciata, Nelson, Tr. Conn. Acad., v. 2, p. 189.	Tert ; Peru.
T. cingulata, Shy.	
Id. Hup, Gay's Chile, p. 154.	Quart.; Chile.
T. cochleiformis, Gabb, J. Conch., 1869, p. 29.	Tert.; Peru.
T. Patagonica, Shy., Darwin, p. 256, Pl. 3, f. 48.	
T. *Chilensis*, Shy , *loc. cit.*, p. 257, Pl. 4, f. 1.	
Cerithium cælatum, Con., P. Acad. Nat. Sci., Phila., 1846, p. 19, Pl. 1, f. 19.	Tert.; Pat.; Chile.
T. plana, Nelson, Tr. Conn. Acad., v. 2, p. 186.	Tert.; Peru.
T. suturalis, Nelson, *id.*, p. 188.	Tert.; Peru.
Crucibulum inerme, Nelson, *id.*, p. 188.	Tert.; Peru.
Trochita trochiformis, Chemn. (sp.).	
Patella id., Chemn., Conch. Cab., p 355, Pl. 168, f. 1626, 1627.	
Trochus radians, Lam., A. S. V., v. 7, p. 11.	
Calyptræa Araucana, Less., Voy Coquille, p. 386.	
C. trochiformis, d'Orb., Moll. A Mer., p. 461, p. 59, f. 3.	
Infundibulum id., d'Orb., Foss. A. M., p. 158.	
Trochita radians, H. & A. Ad. Gen. Moll.	

C. trachiformis, Hup., Gay's Chile, p. 232.   Quart.; Bol., Chile.
CRYPTA DILATATA, Lam. (sp.), H. & A. Ad., Gen. Moll.
Crepidula id., Lam., A. S. V., v. 6, p. 25.
  Id. d'Orb, A. Mer., p. 159.
Calyptræa id., Hup., Gay's Chile, p. 233.   Quart.; Chile.
C. GREGARIA, Sby. (sp.).
Crepidula id., Sby., Darwin, p. 254, Pl. 3, f. 34.   Tert.; Pat.
NERITINA ORTONI, Con., J. Conch., v. 6, p. 195, Pl. 10, f. 11.
N. pupa, Gabb (not Lin.), J. Conch., v. 4, p. 197, Pl. 16, f. 2.   Tert.; Brazil.
?TROCHUS ROUAULTII, Hup., Gay, p. 148.   Quart.; Chile.
MONODONTA PATAGONICA, d'Orb. (sp.).
Trochus id., d'Orb., Moll. A. Mer., p. 408, Pl. 55, f. 1–4.   Quart.; A. Rep., Pat.
CHLOROSTOMA LUCTUOSUM, d'Orb. (sp.), H. & A. Ad., Genera, p. 428.
Trochus id., d'Orb., Moll. A. Mer., p. 409, Pl. 76, f. 16–19.
  Id. d'Orb., Foss. A. Mer., p. 155.   Quart.; Bol.
CALLOPOMA LINEATUM, Nelson, Tr. Conn. Acad., v. 2, p. 186, Pl. 6, f. 2.   Tert.; Peru.
C. NODULIFERUM, Nels., *loc. cit.*, p. 187, Pl. 6, f. 1.   Tert.; Peru.
DENTALIUM GIGANTEUM, Sby., Darwin, p. 263, Pl. 2, f. 1.
D. corrugatum, Hup., Gay, p. 276, Pl. 2, f. 8.   Tert.; Chile.
D. INTERMEDIUM, Hup., Gay, p. 276, Pl. 2, f. 9.   Tert.; Chile.
D. MAGUS, Sby., Darwin, p. 263, Pl. 2, f. 3.   Tert.; Chile.
D. SULCOSUM, Sby., *Id.*, p. 263, Pl. 2, f. 2.   Tert.; Chile.
FISSURELLA CRASSA, Lam., A. S. V., v. 6, p. 11.
  Id. Hup., Gay, p. 240.
  Id. d'Orb., A. Mer., p. 159.   Quart.; Chile, Bol.
SIPHONARIA LESSONI, Blainv. Malac., Pl. 44, f. 2.
  Id. d'Orb., A. Mer., p. 159.   Quart.; Arg. Rep.
PATELLA MEXICANA, Brod. & Sby., Zool. Jour., v. 4, p. 369.
  Id. Rve., Icones, Sp. 1.   Tert.; Peru, (Payta).
I have a remarkably fine example among the Raimondi fossils.
ACMÆA SUBRUGOSA, d'Orb., Moll. A. Mer., p. 497.
  Id. d'Orb., Foss. A. Mer., p. 160.   Quart.; Arg. Rep.
CHITON TUBERCULIFERUS, Sby., Tank. Cat.
C. spiniferus, Frembly, Zool. Jour., p. 196, Pl. 16, f. 1.
C. tuberculiferus, d'Orb., A. Mer., p. 159.   Quart.; Bo.
CHILINA ANTIQUATA, d'Orb., A. Mer., p. 114.   Tert.; Pat.
BULLA AMBIGUA, d'Orb., A. Mer., p. 113, Pl. 12, f. 1–3.
  Id. Hup., Gay's Chile, p. 86.   Tert.; Chile.
B. COSMOPHILA, Sby., Darwin, p. 254, Pl. 3, f. 35.   Tert.; Chile.
BULIMUS LINTEUS, Con., J. Conch., v. 6, p. 195, Pl. 10, f. 9.   Tert.; Brazil.
LIRIS LAQUEATA, Con., *loc. cit.*, p. 194, Pl. 10, f. 3.   Tert.; Brazil.
CYCLOCHEILA PEBASANA, Con., P. Acad. Nat. Sci., Phila., v. 26, p. 32, Pl. 1, f. 17.   Tert.; Brazil.
CIRRODASIS VENUSTA, Con., *loc. cit.*, p. 31, Pl. 1, f. 15.   Tert.; Brazil.

TOXOSOMA EDORKA, Con., *loc. cit.*, p. 31, Pl. 1, f. 7. — Tert.; Brazil.
?PACHYTOMA TERTIANA, Con., *loc. cit.*, p. 31, Pl. 1, f. 11. — Tert.; Brazil.

LAMELLIBRANCHIATA.

DACTYLINA CHILOENSIS, Molina (sp.).
Pholas id., Molina, Hist. Chile, p. 179.
Id. Hup., Gay's Chile, p. 381, Pl. 6, f. 3.
Dactylina id., Gabb, J. Conch., 1869, p. 29. — Tert.; Chile, Peru.
SOLECURTUS HANETIANA, d'Orb.
Solenocurtus id., d'Orb., A. Mer., p. 124, Pl. 15, f. 1–2.
Solecurtus id., Hup., Gay's Chile, p. 368. — Tert.; Chile.
CORBULA BRADLYI, Nelson, Tr. C. Acad., v. 2, p. 200. — Tert.; Peru.
AZARA LABIATA, Matton (sp.), d'Orb., A. Mer., p. 161.
Mya id., Matton. — Quart.; Arg. Rep.
PACHYDON ALATUS, Con., J. Conch., v. 6, p. 197, Pl. 11, f. 1.
Id. Con., P. Acad. Nat. Sci., Phila., v. 26, p. 28, Pl. 1, f. 4, 18. — Tert.; Brazil
P. CARINATUS, Con., J. Conch., v. 6, p. 196, Pl. 10, f. 7.
Anisothyris id., Woodw., Ann. M. N. H., 1871, p. 108, Pl. 5, f. 6. — Tert.; Brazil.
P. CUNEATUS, Con., J. Conch., v. 6, p. 179, Pl. 10, f. 12.
Anisothyris id., Woodw., *loc. cit.*, p. 107, Pl. 5, f. 8–a–b.
P. id., Con., Proc. Acad. Nat. Sci., Phila., v. 26, p. 28, Pl. 1, f. 3. — Tert.; Brazil.
P. (*Anisorhyncus*) CUNEIFORMIS, Con., *loc. cit.*, p. 28, Pl. 1, f. 19. — Tert.; Brazil.
P. (*A.?*) DISPAR, Con., *loc. cit.*, p. 27, Pl. 1, f. 1. — Tert.; Brazil.
P. ERECTUS, Con., J. Conch., v. 6, p. 197, Pl. 10, f. 16.
Anisothyris id., Woodw., A. M. N. H., 1871, p. 107, Pl. 5, f. 9–a–b.
P. id., Con., Proc. Acad. Nat. Sci., Phila., v. 26, p. 28. — Tert.; Brazil.
P. LEDÆFORMIS, Dall (sp.).
Corbula (Anisothyris?) id., Dall, J. Conch., v. 8, p. 92, Pl. 16, f. 14–15. — Tert.; Brazil.
P. OBLIQUUS, Gabb, Jour. Conch., v. 4, p. 199, Pl. 16, f. 5.
Id. Con., J. Conch., v. 6, p. 197, Pl. 10, f. 15.
Anisothyris id., Woodw., A. M. N. H., 1871, p. 106, Pl. 5, f. 5–a–b. — Tert.; Brazil.
P. OVATUS, Con., J. Conch., v. 6, p. 197, Pl. 10, f. 4.
Anisothyris id., Woodw., A. M. N. H., 1871, p. 106. — Tert.; Brazil.
P. TENUIS,* Gabb, J. Conch., v. 4, p. 199, Pl. 16, f. 6.
P. id., Con., J. Conch., v. 6, p. 196, Pl. 10, f. 1.
Anisothyris Hauxwelli, Woodw., A. M. N. H., 1871, p. 105, Pl. 5, f. 7–a–d.
Tellina Amazonensis, Gabb, J. Conch., v. 4, p. 198, Pl. 16, f. 4. — Tert.; Brazil.

* In insisting on the use of this name, I am sustained by all the rules of scientific nomenclature. Otherwise I should be very loath to raise a question over so paltry a matter as the privilege of claiming a genus or species. Mr. Conrad proposed *Anisothyris* because, as he says, he was requested to do so, *Pachydon* being considered objectionable; at the same time he described all of his own species under the old generic name, and subsequently retracted *Anisothyris* and described several more, also, as *Pachydon*. Woodward arbitrarily

MADE BY DR. ANTONIO RAIMONDI IN PERU. 333

?HOMOMYA RUGATA, Sby. (sp.).
Mactra? id., Sby., Darwin, p. 249, Pl. 2, f. 8. Tert.; Pat.
MACTRA COLCHAGUANA, Hup., Gay, p. 349. Tert.; Chile.
M. ZORRITENSIS, Nelson, Tr. Conn. Acad., v. 2, p. 201. Tert.; Peru.
STANDELLA AUCA, d'Orb. (sp.).
Mactra id., d'Orb., A. Mer., p. 125, Pl. 14, f. 19–20.
    Id.   Hup., Gay's Chile, p. 349. Tert.; Chile.
?HEMIMACTRA DARWINII, Sby. (sp.).
Mactra id., Sby., Darwin, p. 249, Pl. 2, f. 9. Tert.; Pat.
LUTRARIA PLICATELLA, Lam., A. S. V., v. 5, p. 470.
    Id.   d'Orb., A. Mer., p. 161. Quart.; Pat.
RÆTA GIBBOSA, Gabb, J. Conch., v. 4, p. 30.
    Id.   Gabb, present paper. Tert.; Peru.
OSTOMYA PAPYRIA, Con., Proc. Acad. Nat. Sci., Phila., v. 26, p. 30, Pl. 1, f. 6. Tert.; Brazil.
PERONÆA PETITIANA, d'Orb. (sp.).
Tellina id., d'Orb., Moll. A. Mer., p. 537, Pl. 81, f. 26–27.
Tellinides? oblonga, Sby., Darwin, p. 250, Pl. 2, f. 12. Tert.; Chile.
MACOMA HUAFFOENSIS, Con. (sp.).
Tellina id., Con., Proc. Acad. Nat. Sci., Phila., 1846, p. 20, Pl. 1, f. 20. Tert.; Pat.
AMPHIDESMA BREVIROSTRUM, Hup., Gay, p. 361, Pl. 6, f. 1. Tert.; Chile.
STRIGILLA PRORA, Hanley (sp.).
Tellina id., Hanley, Zool. Proc., 1844.
Strigilla id, Gabb, J. Conch., 1869, p. 30. Tert.; Peru.
MESODESMA DONACINA, Rve., Conch., Pl. 45, f. 1.
Donacilla Chilensis, d'Orb., Moll. A. Mer., p. 530. Quart.; Chile.
VENUS BAYLII, Hup., Gay, p. 340. Tert.; Chile.
?V. CHILENSIS, d'Orb., A. Mer., p. 124, Pl. 13, f. 12–13.
Lucina id., d'Orb., on plate.
V. id., Hup., Gay's Chile, p. 342. Tert.; Chile.
?V. CLERYANA, d'Orb., A. Mer., p. 123, Pl. 13, f. 7–8.
    Id.   Hup., Gay, p. 341. Tert.; Chile.
?V. COQUANDI, Hup., Gay, p. 340. Tert.; Chile.
V. DOMBEI, Lam., A. S. V., v. 5, p. 590.

brushes aside both generic and specific name, saying, in the former case, "the objection to *Pachydon* is too obvious to need any further delay in condemning it." Mr. Conrad's objection is that its derivation is the same as *Pachyodon* (=? *Pazyodon* Schum.) Woodward's objection to the specific name—*tenuis*—is that his specimens were *not* thin ; mine were ; and there is no rule of nomenclature that authorizes one writer to change a name once given to a species, by another, unless it is pre-occupied. As to the name *Pachydon*, there are numerous precedents for the elision of an uneuphonic letter in generic names ; and if, at the worst, the word should have no meaning ; if it is a "nonsense name," there are not wanting precedents—and eminent ones. Mr. Woodward would not presume to amend Adanson and others ! The etymology of *Pachydon* is presumed to be the same as that of Schumacher's *Pizyodon*, but there can be no possible chance of confusion between the two words, either for the eye or ear. I was not ignorant of the existence of this name when I proposed my own ; nor yet of Stutchbury's *Pachyodon*, which is synonymous with *Thalassides* of Berger.

70

834   DESCRIPTION OF A COLLECTION OF FOSSILS,

V. Dombei, d'Orb., A. Mer., p. 160.	Quart.; Bol.
V. Hanetiana, d'Orb., A. Mer., p. 123, Pl. 13, f. 3-4.	
Id. Hup., Gay, p. 341.	Tert.; Chile.
V. opaca, Brod., Zool. Proc.	
Id. d'Orb., A. Mer., p. 160.	Quart.; Bol.
?V. Petitiana, d'Orb., A. Mer., p. 123, Pl. 13, f. 9-11.	
Id. Hup., Gay's Chile, p. 342.	Tert.; Chile.
?V. pulvinata, Hup., Gay, p. 343.	Tert.; Chile.
Chione lenticularis, Sby. (sp.), H. & A. Ad., Gen.	
Venus id., Sby., Proc. Zool. Soc., 1835, p. 42.	
Id. Hup., Gay, p. 336, Pl. 6, f. 1.	Quart.; Chile.
C. meridionalis, Sby. (sp.).	
Venus id., Sby., Darwin, p. 250, Pl. 2, f. 13.	Tert.; Chile; Pat.
C. Munsteri, d'Orb. (sp.).	
Venus id., d'Orb., A. Mer., p. 121, Pl. 7, f. 10-11.	Tert.; Pat.; Arg. Rep.
?C. Patagonica, d'Orb. (sp.).	
Venus id., d'Orb., A. Mer., p. 160.	Quart.; Pet.
C. subalbicans, Hup. (sp.).	
Venus id., Hup., Gay's Chile, p. 339.	Tert.; Chile.
C. variabilis, Nelson, Tr. Conn. Acad., v. 2, p. 202.	Tert.; Peru.
Callista Rouaultii, Hup. (sp.).	
Venus id., Hup., Gay's Chile, p. 339.	Tert.; Chile.
C. sulculosa, Sby. (sp.).	
Cytherea id., Sby., Darwin, p. 250, Pl. 2, f. 14.	Tert.; Chile.
Dosinia grandis, Nelson, Tr. Conn. Acad., v. 2, p. 201.	Tert.; Peru.
Petricola Chiloensis, Phil., Wiegmanu's Archiv., No. 3.	
Id. Hup., Gay, p. 345.	Quart.; Chile.
Cardium (Cerastoderma) multiradiatum, Sby.	
C. multiradiatum, Sby., Darwin, p. 251, Pl. 2, f. 16.	Tert.; Chile.
Not id. Gabb, Cret. U. States.	
C. (C.) Platense, d'Orb.	
C. Platense, d'Orb., A. Mer., p. 120, Pl. 4, f. 12-14.	Tert.; Arg. Rep.
C. (C.) Puelchum, Sby.	
C. id., Sby., Darwin, p. 251, Pl. 2, f. 15.	Tert.; Chile.
C. (Lævicardium) pertenue, Gabb, Jour. Conch., 1869, p. 30.	
Id. Gabb, present paper.	Tert.; Peru.
C. (Hemicardia) affinis, Nelson.	
Hemicardia id., Nelson, Tr. Conn. Acad., v. 2, p. 204.	Tert.; Peru.
Lucina Patagonica, d'Orb., A. Mer., p. 161.	Tert.; Pat.
?Corbis lævigata, Sby., Darwin, p. 250, Pl. 2, f. 11.	Tert.; Chile.
Crassatella gibbosa, Sby., Proc. Zool. Soc., 1832, p. 56.	
Id. Nelson, Tr. Conn. Acad., v. 2, p. 203.	Tert.; Peru.
C. Lyellii, Sby., Darwin, p. 249, Pl. 2, f. 10.	Tert.; Chile.
Venericardia Patagonica, Sby. (sp.)	

Cardita id., Sby., Darwin, p. 251, Pl. 2, f. 17.	Tert.; Chile.
UNIO DILUVII, d'Orb., A. Mer., p. 127, Pl. 7, f. 12–13.	Tert.; Pat.
ANODON BATESII, Woodw., A. M. N. H., 1871, p. 103, Pl. 5, f. 10.	Tert.; Brazil.
A. PEBASANA, Con., Proc. Acad. Nat. Sci., Phila., v. 26, p. 29, Pl. 1, f. 5.	Tert.; Brazil.
HYRIA LONGULA, Con. (sp.).	
Triquetra id., Con., Proc. Acad. Nat. Sci., Phila., v. 26, p. 29, Pl. 1, f. 10.	Tert.; Brazil.
MYTILUS EDULIFORMIS, d'Orb., A. Mer., p. 162.	Quart.; Arg. Rep.
M. UNGULATUS, L., Gmel., Syst. No. 12.	
Id. Gabb, Jour. Conch., 1869, p. 31.	Tert.; Peru.
DREISSINA (*Mytiloides*) SCRIPTA, Con., P. Acad., v. 26, p. 29, Pl. 1, f. 12–16.	Tert.; Brazil.
PERNA CHILENSIS, Con., Astr. Exped., p. 285, Pl. 48, f. 7.	Tert.; Chile.
P. GAUDICHAUDII, d'Orb., A. Mer., p. 181, Pl. 15, f. 14–16.	Tert.; Chile.
Arca (*Anomalocardia**) BONPLANDIANA, d'Orb.	
Arca id., d'Orb., A. Mer., p. 130, Pl. 14, f. 15–18.	Tert; Pat.; Arg. Rep.
A. (*A.*) LARKENI, Nelson.	
Arca Larkeni, N., Trans. Conn. Acad., v. 2, p. 204, Pl. 7, f. 5–7.	Tert.; Peru.
A. (*Scapharca*) RAIMONDII, Gabb, Jour. Conch., 1869, p. 31.	
Id. Gabb, present paper.	Tert.; Peru.
CUCULLÆA ALTA, Sby., Darwin's S. A., p. 252, Pl. 2, f. 22–23.	Tert.; Pat.
AXINÆA COLCHAGUENSIS, Hup. (sp.).	
Pectunculus id., Huppé, Gay's Chile, p. 302.	Tert.; Chile.
A. PAYTENSIS, d'Orb. (sp.).	
Pectunculus id., d'Orb., A. Mer., p. 129, Pl. 15, f. 11–13.	
Axinæa id., Gabb, Jour. Conch., 1869, p. 31.	Tert.; Peru.
LIMOPSIS INSOLITA, Sby. (sp.).	
Trigonocelia id., Sby., Darwin's S. A., p. 252, Pl. 2, f. 20–21.	Tert.; Pat.
NUCULA PULCHRA, d'Orb., A. Mer., p. 162.	Quart.; Pat.
?N. OLABRA, Sby., Darwin, p. 251, Pl. 2, f. 18.	Tert.; Pat.

This looks much more like one of the *Tellinidæ*, and Sowerby described it only from the surface, not having seen any of the internal characters.

NUCULANA ACUMINATA, Nelson (sp.).

Leda id., Nelson, Tr. Conn. Acad., v. 2, p. 205, Pl. 7, f. 8.	Tert.; Peru.
N. ELEGANS, Hup. (sp.).	
Nucula id., Hup., Gay's Chile, p. 305, Pl. 5, f. 3.	Tert.; Chile.
NUCULANA (*Adrana*) LANCEOLATA, Sby. (sp.), H. & A. Ad., Gen. Moll.	
Nucula id., Sby., Genera, f. 1.	
Id. d'Orb., A. Mer., p. 162.	Quart.; Pat.

---

* The name *Anomalocardia* given by Klein to this group of *Arcas* was undoubtedly the oldest, and it has been adopted by some modern authors, so that no subsequent synonym has been proposed. As I have expressed myself emphatically elsewhere against the use of Klein's names, there is little to add. He was only occasionally, and by accident binomial. I admit the name here, however, to avoid the necessity of proposing a new one myself. Schumacher's *Anomalocardia*, now half a century old, is valid as a genus, or subgenus (according to individual views) in the *Veneridæ*, and here is a good opening for somebody, more ambitious than myself, to attach his name to a good subgenus, or genus, as he may prefer to call it, in the *Arcas*.

336   DESCRIPTION OF A COLLECTION OF FOSSILS.

NEILO ORNATA, Sby. (sp.).
Nucula id., Sby., Darwin's S. A., p. 251, Pl. 2, f. 19.   Tert.; Pat.
PECTEN ACTINODES, Sby., *id.*, p. 253, Pl. 3, f. 33.   Tert.; Pat.
P. CENTRALIS, Sby., *id.*, p. 253, Pl. 3. f. 31.   Tert.; Pat.
P. GEMINATUS, Sby., *id.*, p. 252, Pl. 2, f. 24.   Tert.; Pat.
P. PARANENSIS, d'Orb., A. Mer., p. 132, Pl. 7, f. 5–9.
  Id.   Sby., Darwin's S. A., p. 253, Pl. 3, f. 30.
Very closely allied to *P. Madisonius* of the North American Miocene, but differs in having a few more ribs, although the details of their ornaments are the same. Mr. Conrad, whose experience with these fossils extends over nearly half a century, and therefore must be treated with great deference, assures me that the North American fossils differ more among themselves, than does the South American form from the typical *Madisonius*, and that he would not hesitate in uniting them.

P. PATAGONENSIS, d'Orb., A. Mer., p. 131, Pl. 7, f. 1–4.   Tert.; Pat.
P. PROPINQUUS, Hup., Gay's Chile, p. 291, Pl. 5, f. 2.   Tert.; Chile.
P. PURPURATUS, Lam., A. S. V., v. 7, p. 134.
  Id.   Gabb, J. Conch., 1869, p. 32.   Tert.; Peru.
P. RUDIS, Sby., Darwin's S. A., p. 254, Pl. 3, f. 32.   Tert.; Chile.
P. TENUICOSTATUS, Hup., Gay, p. 291, Pl. 5, f. 4.   Tert.; Chile.
P. (*Pleuronectia*) DARWINIANUS, d'Orb.
Pecten Darwinianus, d'Orb., A. Mer., p. 133.
  Id.   Sby., Darwin's S. A., p. 253, Pl. 3, f. 28–39.   Tert.; Pat.; A. Rep.
ANOMIA ALTERNANS, Sby., *id.*, p. 252, Pl. 2, f. 25.   Tert.; Chile.
OSTREA ALVAREZII, d'Orb., A. Mer., p. 134, Pl. 7, f. 19.   Tert.; Arg. Rep.
O. COPIAPINA, Con., U. S. Naval Astron. Exped., p. 285.   Tert.; Chile.
O. FERRARISI, d'Orb., A. Mer., p. 134, Pl. 7, f. 17–18.   Tert.; Pat.
O. GALLUS, Val., Voy. Venus, Pl. 21, f. 1.
O. *Cerrosensis*, Gabb, Pat. Col., v. 2, p. 35, Pl. 11, f. 61.
O. *gallus*, Gabb, J. Conch., 1869, p. 32.   Tert.; Chile.
O. MAXIMA, Hup., Gay, p. 282, Pl. 4, f. 1.   Tert.; Chile.
O. PATAGONICA, d'Orb., A. Mer., p. 133, Pl. 7, f. 14–16.   Tert.; Pat.; Arg. Rep.
O. PUELCHANA, d'Orb., A. Mer., p. 162.   Quart.; Arg. Rep.
O. ROSTRATA, Hup., Gay's Chile, p. 283.   Tert.; Chile.
O. TRANSITORIA, Hup., Gay's Chile, p. 283, Pl. 4, f. 3.   Tert.; Chile.

BRACHIOPODA.

TEREBRATULA FONTAINEI, d'Orb., A. Mer.
T. Chilensis, d'Orb., Foss. A. Mer., p. 163.
Not *T. id.*, Brod.
T. Fontainei, Hup., Gay's Chile, p. 400.   Quart.; Chile.
T. PATAGONICA, Sby., Darwin's S. A., p. 252, Pl. 2, f. 26–27.   Tert.; Pat.

RADIATA.
ECHINODERMATA.

ECHINUS PATAGONENSIS, d'Orb., A. Mer., p. 135, Pl. 6, f. 14–16.   Tert.; Pat., Arg. Rep.

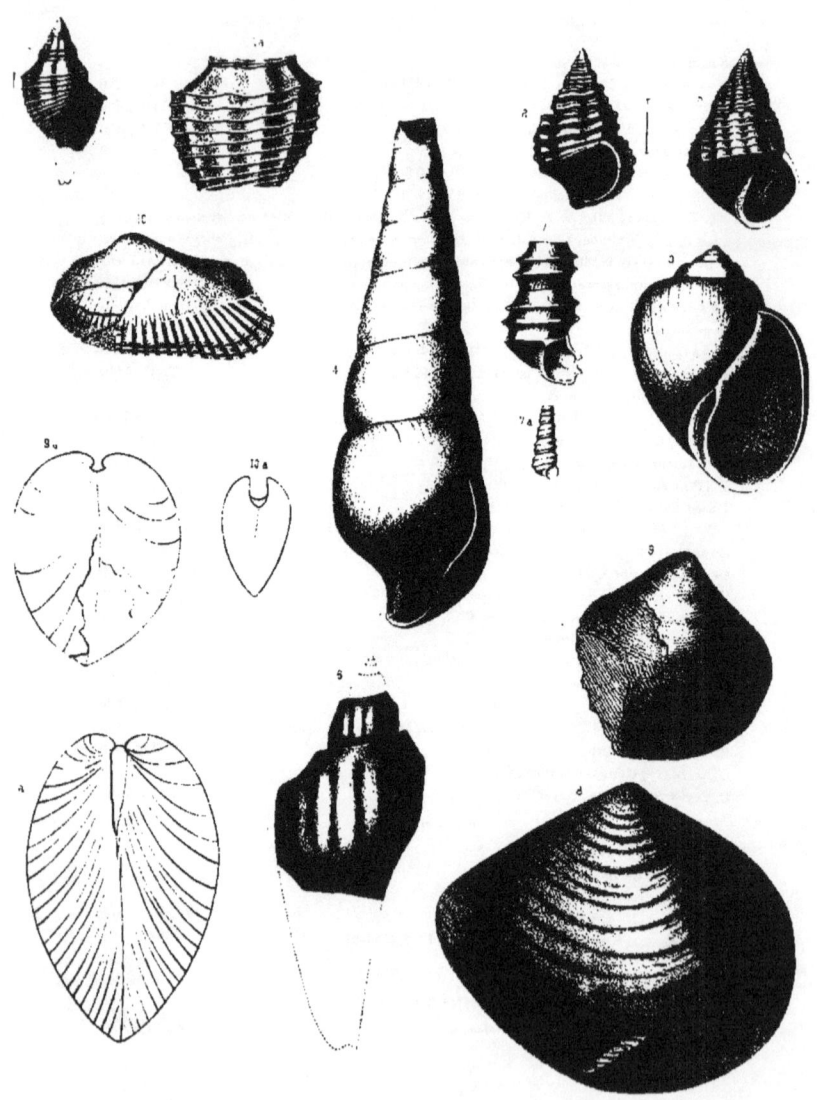

Gabb on Peruvian Fossils.

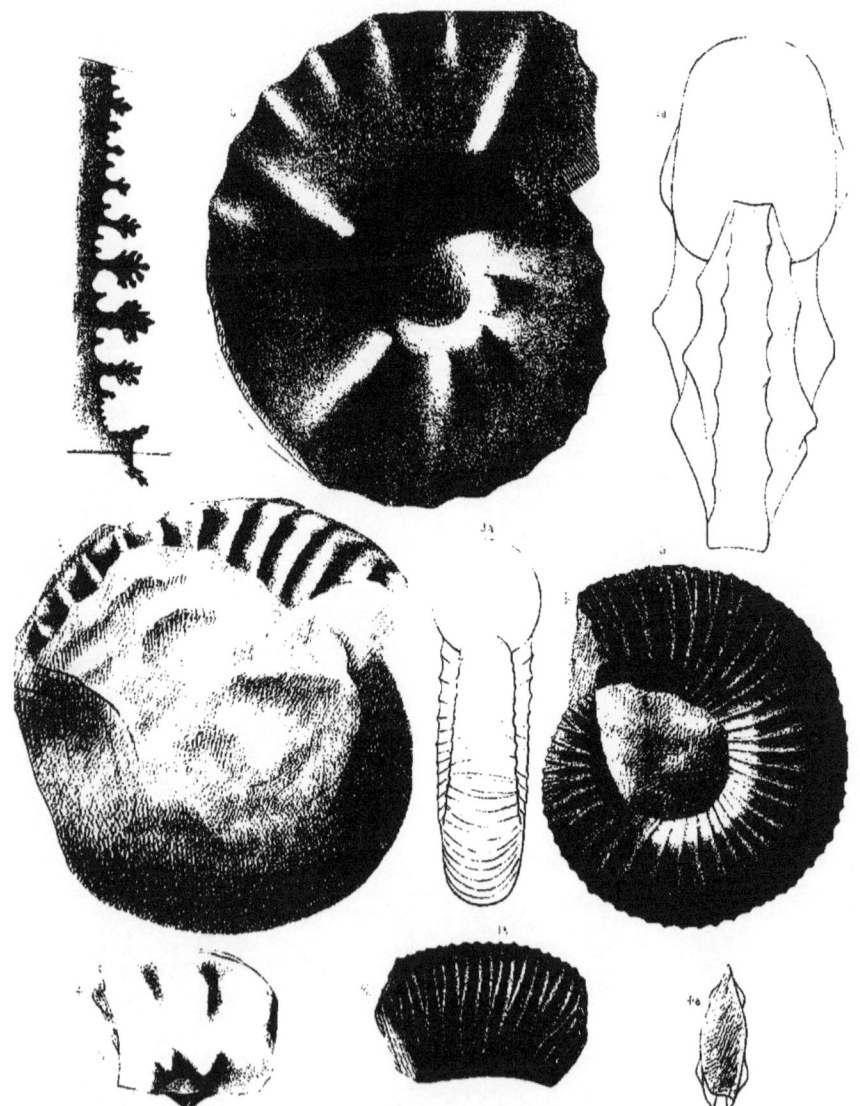

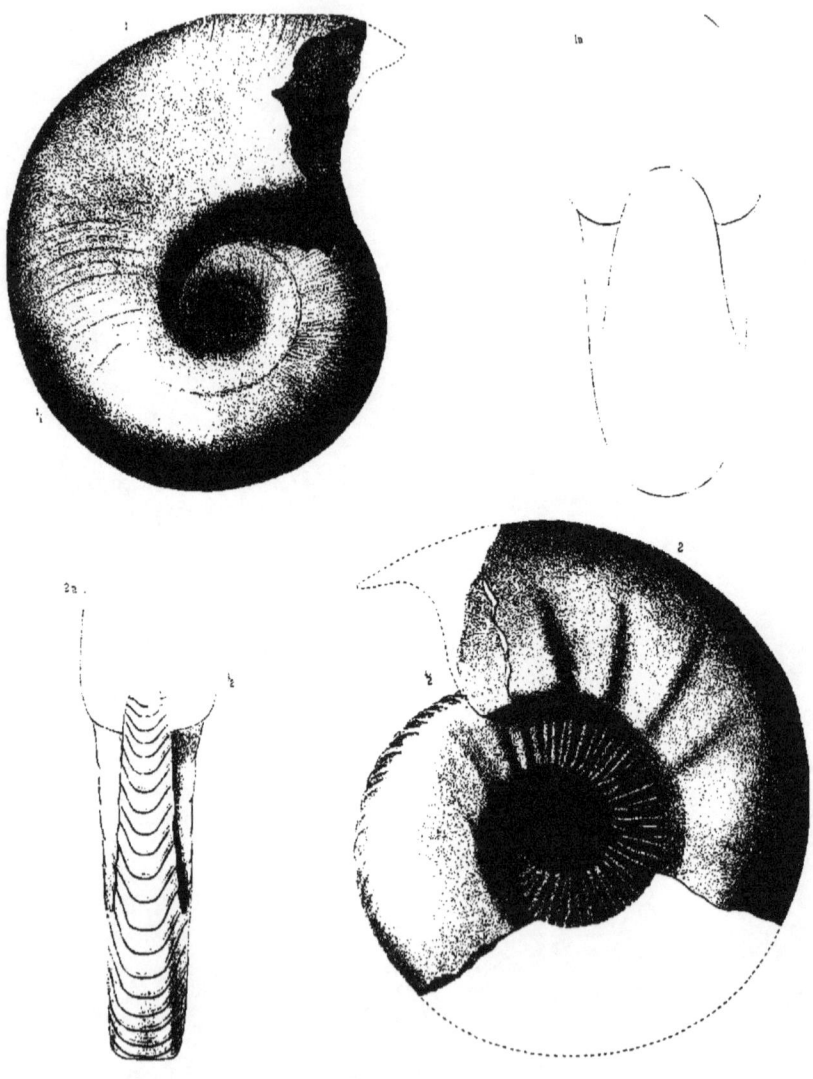

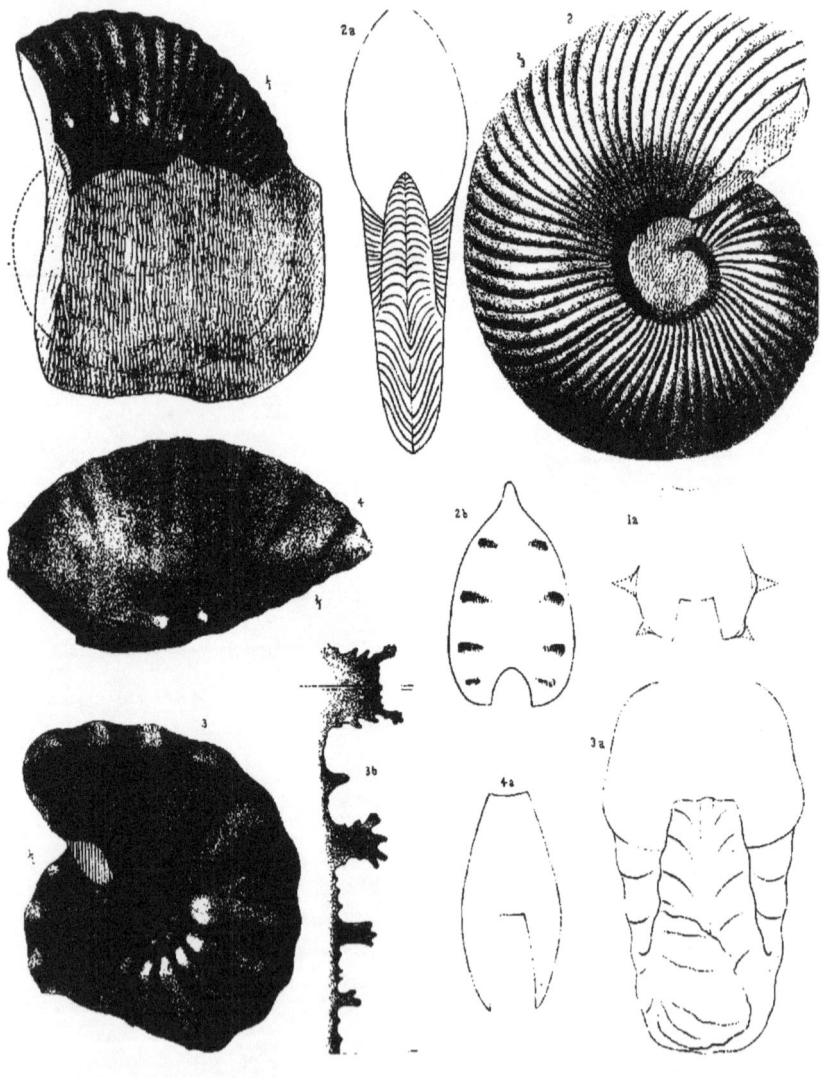

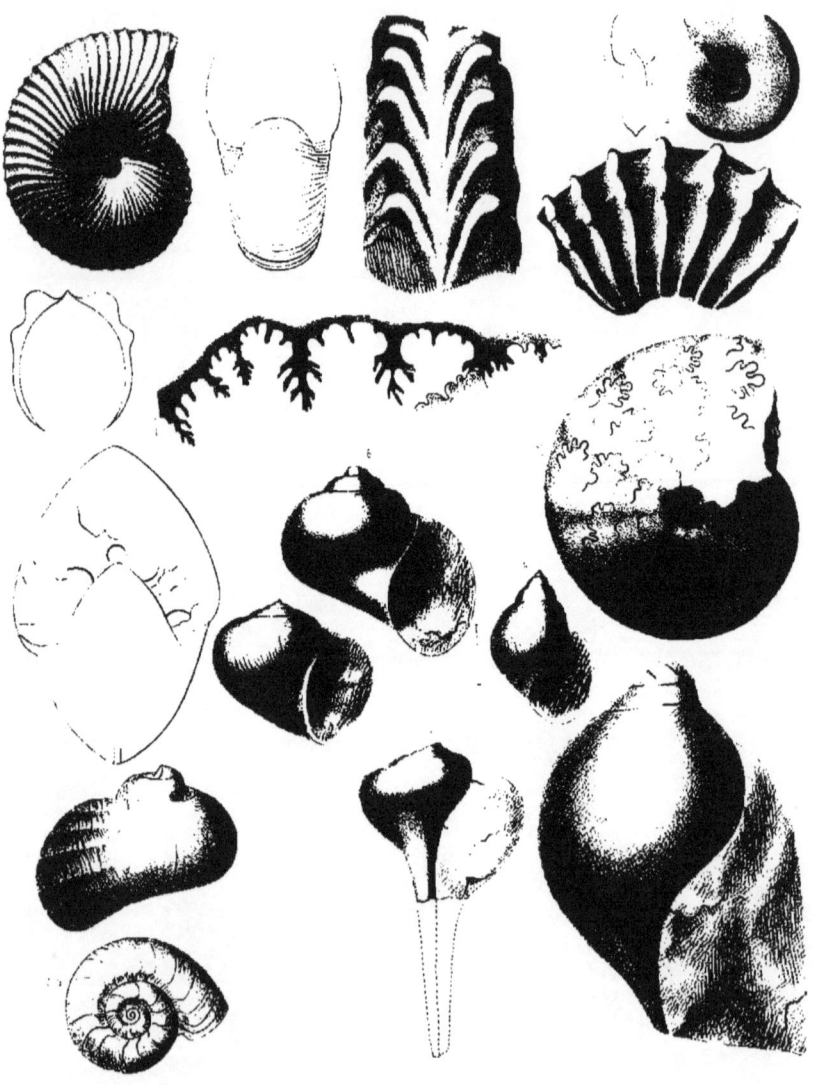

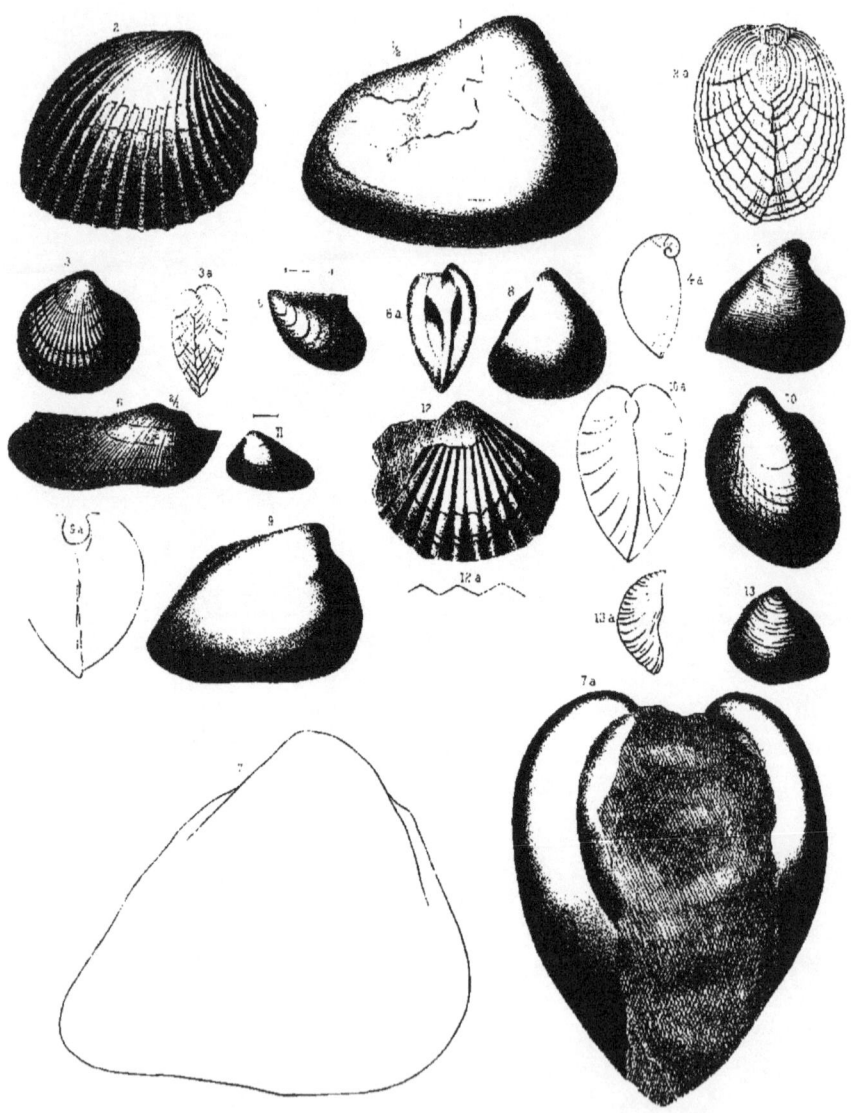

Gabb on Peruvian Fossils.

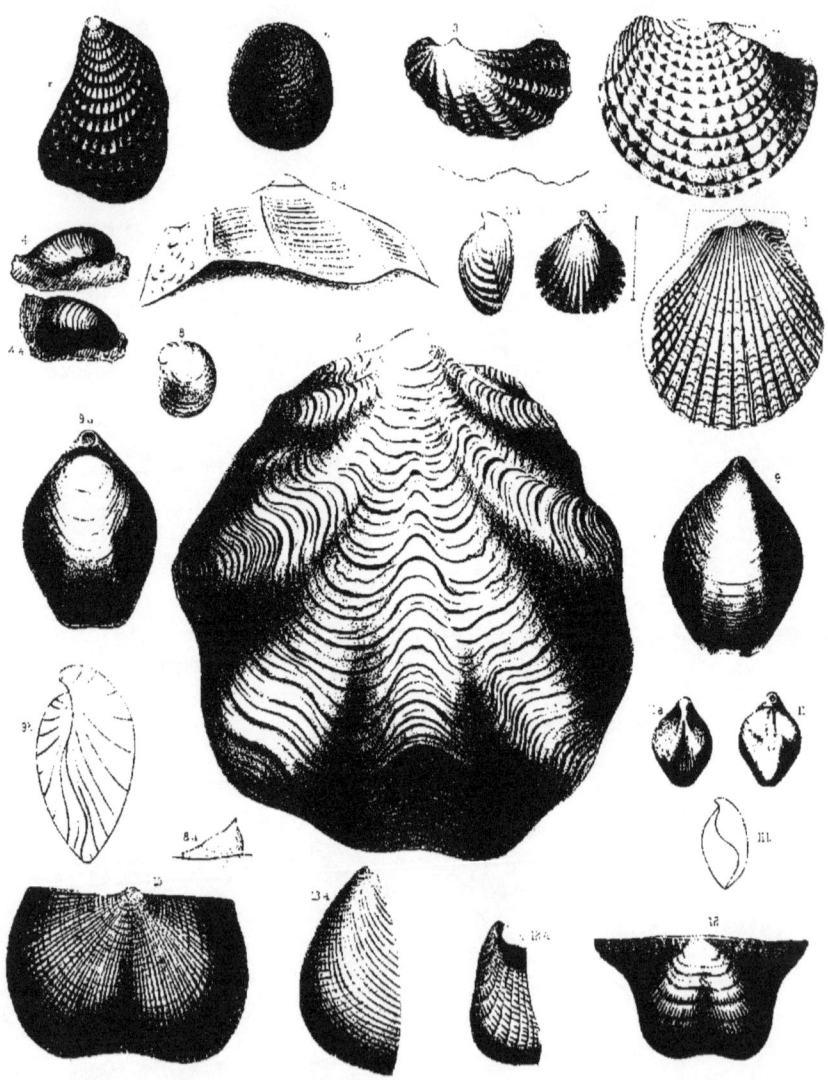

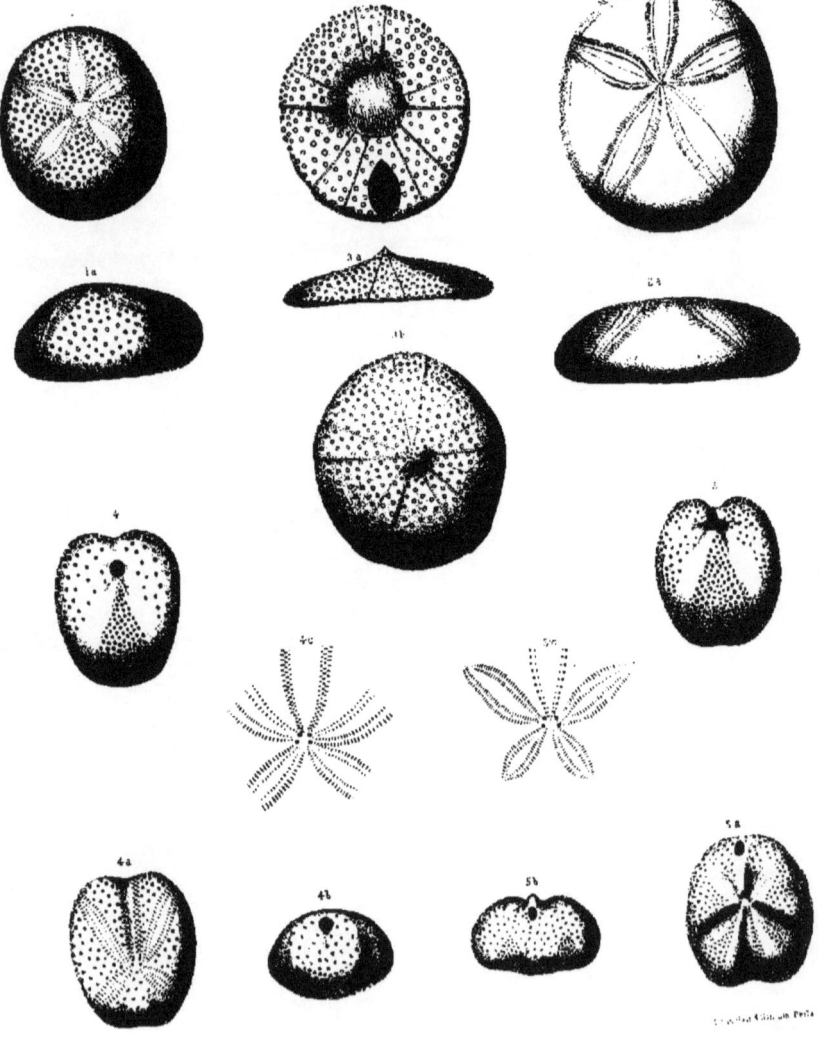

www.ingramcontent.com/pod-product-compliance
Lightning Source LLC
Chambersburg PA
CBHW032243080426
42735CB00008B/983